John Matthew Jones

THE NATURALIST IN BERMUDA

Elibron Classics
www.elibron.com

Elibron Classics series.

ISBN 1-4021-8071-3 (paperback)
ISBN 1-4021-1169-X (hardcover)

This Elibron Classics Replica Edition is an unabridged facsimile of the edition published in 1859 by Reeves & Turner, London.

THE NATURALIST IN BERMUDA.

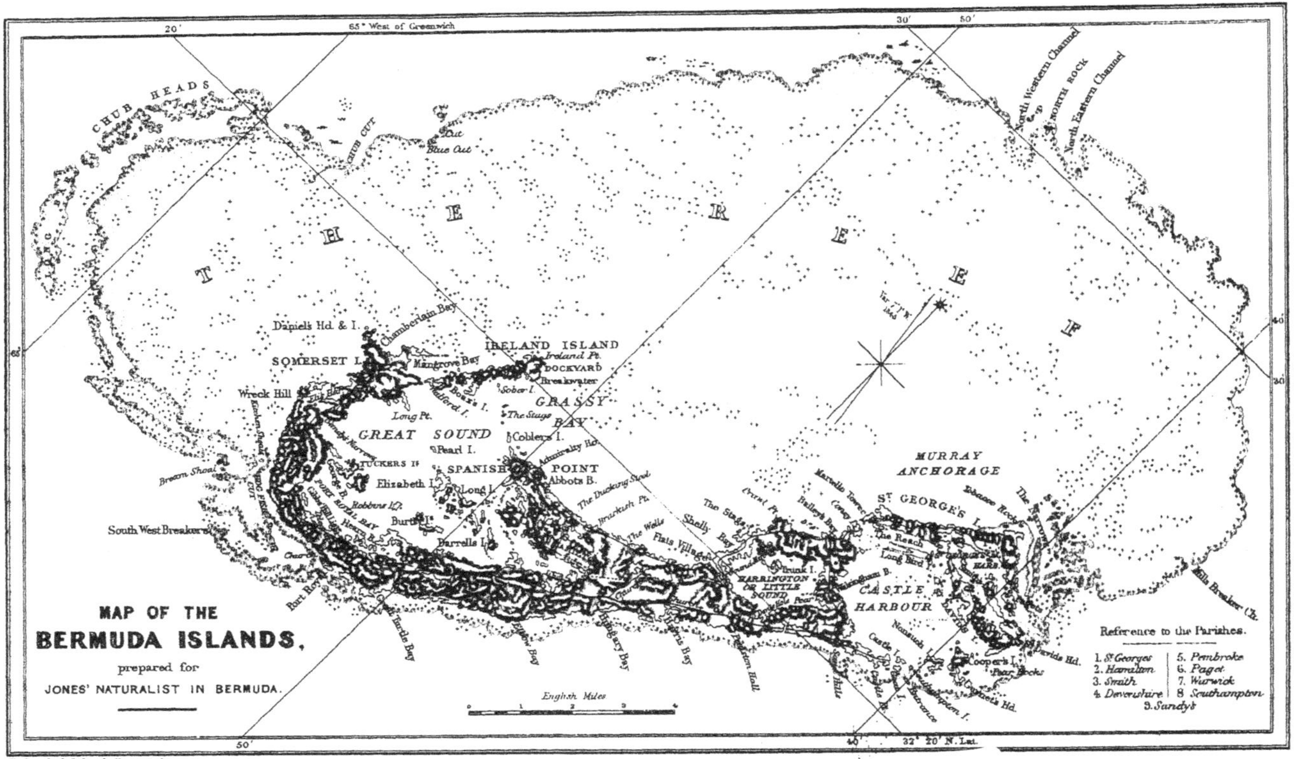

E. Stanford, Litho. 6, Charing Cross.

THE

NATURALIST IN BERMUDA;

A SKETCH OF THE

GEOLOGY, ZOOLOGY, AND BOTANY,

OF THAT REMARKABLE GROUP OF ISLANDS;

TOGETHER WITH

METEOROLOGICAL OBSERVATIONS.

BY

JOHN MATTHEW JONES, ESQ.,

(Of the Middle Temple.)

ASSISTED BY

MAJOR J. W. WEDDERBURN (*Late 42nd Roy. Highlanders*), AND J. L. HURDIS, ESQ.

WITH A MAP AND ILLUSTRATIONS.

"Every kingdom, every province, should have its own monographer."

GILBERT WHITE.

LONDON:

REEVES & TURNER, 238, STRAND.

1859.

PREFACE.

In publishing this little book, containing the first account ever submitted to the public of the Natural History of the Bermudas, the author trusts that its numerous and necessary imperfections will meet with the lenient consideration of his readers, and more particularly of learned naturalists, who are requested to regard this first effort as a small and humble contribution to the present state of natural knowledge, and as a prelude to a more complete publication on the same subject.

The author has to return his thanks to many kind friends for their generous endeavours to aid the progress of the work; more especially to Major Wedderburn and Mr. Hurdis, who laid their notes and observations, made during many years' residence on the Islands, entirely at his disposal, and have otherwise greatly assisted in the formation of the volume.

The author has also to express his acknowledgements—to Colonel Nelson, of the Royal Engineers, for the information contained in the Geological paper; to John H. Trott, Esq., Provost Marshall-

General of Bermuda, for unceasing efforts to aid the author in his researches during his residence in the Bermudas; to R. S. Wood, Esq., of Walsingham, and W. B, Smith, Esq., of Mangrove Bay, for notes of several species; and to Miss L. L. Deudney, Miss Ella Tucker, and Mr. Hurdis, for sketches which have afforded the vignettes for this volume.

The oval views, at the commencement of each division, are from photographs taken upon the Islands by Mr. Whittemore, of New York.

CONTENTS.

ILLUSTRATIONS.

INTRODUCTION.

THE small, yet interesting group of Islands, known as the "Bermudas," is situate in 32° 15′ North Latitude, and 64° 51′ West Longitude, and distant from Cape Hatteras, the nearest point of the North American Coast, about six hundred nautical miles.

These Islands were discovered by Juan Bermudez, in a Spanish vessel named "La Garza," bound from Spain to the Island of Cuba, and having on board Gonzalez Oviedo, the celebrated historian of the Indies. Having approached within cannon-shot range of the southern shore, the discoverers, judging from the appearance of the land, regarded the group as a single island, about twelve leagues in length and thirty in circumference; they also concluded that it was uninhabited by man, and resolved to send boats on shore to make observations, and leave a few hogs, which might breed and be afterwards useful. When, however, they were preparing to disembark, a strong contrary gale arose, which obliged them to sheer off, and be content with the view already obtained. Oviedo calls it "the remotest island in the whole world," meaning, we presume, the most distant from any land, and mentions the swarms of birds and flying-fish, with the contests between them, as presenting one of the most amusing spectacles he had ever beheld.*

The first native of England known to have set foot upon the Islands, was a mariner named Henry May, who, while on a voyage from the West Indies to Europe, in a French vessel, in the year 1593, was wrecked upon the north-west reefs, several miles distant from the shore. He found the land overgrown with trees of various kinds, though chiefly with the cedar; "many hogs" were also met with, but these were so lean as to be unfit for food; birds, fish, and turtle were in great abundance. Fortunately for May and his French companions, the carpenter's tools, with a portion of the sails and rigging of the ship, were saved by them before the wreck went to pieces. This enabled them to cut down cedars and construct "a barque of eighty tunnes," in which, after a sojourn of nearly five months in those Isles, they all set sail on the 11th May. On the 20th of the same month they made the Island of Cape Breton, when they took in wood and water, and sailed for the banks of Newfoundland. Here they met with many ships, but none

* Murray's British America.

of them charitably inclined towards them, when it pleased God they fell in with "the honest English barque Fawmouth," which received them on board. While with this vessel they "tooke" a French ship, into which May's dear friend Captain de la Barbotier, and his seamen, were transferred; May himself remaining with the English vessel, which arrived at Falmouth, in August, 1594.

The next published account of a visit to the Bermudas is contained in an old black letter work, entitled, "The Wreck of the Sea Adventure," by Sil. Jourdan, a copy of which scarce work is in the library of the British Museum.

The "Sea Adventure" was one of a small fleet of ships which sailed from England in the year 1609, for the Colony of Virginia, having on board Sir Thomas Gates, the newly appointed Governor of that possession, Admiral Sir George Somers,* and other persons, beside the crews. After describing minutely the horrors of a terrific storm, which separated the "Sea Adventure" from the rest of the fleet, and drove her, in a shattered condition, upon the reefs of Bermuda, where she became a total wreck, the writer proceeds to describe the natural features of those islands.

Weeds and plants of several kinds; tall and goodly cedars, with "infinite store" of palmettos, mulberries, wild olive, and other trees, where found everywhere. Sea-birds were particularly abundant, and evinced that absence of fear towards man, which even at the present day, is noticed by navigators and others, when visiting isolated rocks or uninhabited islands. Fish of many kinds were obtained in large quantities, and required little piscatory skill or fineness

* The Bermudas, in former times, were better known as the "Somers' Isles," a title frequently corrupted into "Summer Isles."

of tackle in their capture; indeed, the writer amusingly states that, "if a man step into the water they will come round about him, so that men were faine to get out for fear of byting." Hogs, supposed to have been introduced by Bermudez, or by some subsequent and unrecorded navigator, had increased so largely, as to enable Sir George Somers to kill thirty-two in a single day's hunt.

In May, 1610, Sir George Somers and his companions embarked in two small vessels which they had themselves constructed, and sailed for the settlement of James Town, in Virginia, where they arrived in safety. Here Sir George found the settlers in much distress for want of supplies, and, although upwards of sixty years of age, gallantly volunteered to return to the Bermudas, in his little cedar-built craft of thirty tons, for the purpose of obtaining hogs. It was on this expedition, and on the site where the town of Saint George now stands and bears his name, that this worthy admiral breathed his last, exhorting his companions to return with all diligence to Virginia.

Captain Matthew Somers, the nephew and heir of Sir George, who appears to have inherited the gallant spirit of his uncle, in place of returning to Virginia, formed the daring resolution of navigating the same small cedar-built vessel to England, taking with him the embalmed remains of his departed relative. Fortune smiled upon the bold undertaking, and the ship arrived, in due time, safely at Whitchurch, in Dorsetshire, where the admiral's remains were buried with military honors, and an epitaph, in Latin, inscribed upon his tomb.

Upon the representation made by Captain Somers and his companions, a company was formed in England for the purpose of colonizing the Bermudas, and a ship with emigrant settlers, under the charge of Governor Richard More, having arrived in those islands on the 11th July, 1612, they may be considered as permanently inhabited by the human race from that period.

Although the Bermudas have undergone some change since the days of Sir George Somers, and a portion of the land has been brought under cultivation, still we are inclined to look upon the general outline and appearance of those islands at the present time, as very similar to that which presented itself to the gallant old admiral. The cedar tree still clothes the uncultivated hills and valleys with its evergreen foliage, and the palmetto still dots the landscape with its bending plumes. Fish still abound in the surrounding waters; but the wild hog has long since given place to the domestic representative of the same family. The tropic bird and the tern still frequent the rocky coast during the fervid months of summer, for the purpose of incubation, disappearing at the approach of winter, but the great family of gulls and other sea-birds, which tends so much to the beauty and cheerfulness of ocean scenery, has long since abandoned a spot so thickly inhabited by the human race.

That the Bermudas afford an excellent position from whence to observe the annual migration of many species of the feathered tribes of America, cannot be doubted. Equidistant, or nearly so, from the shores of Nova Scotia, the United States, and the West Indian archipelago, they present, as it were, a casual resting place to many birds, while traversing the broad expanse of ocean which forms the eastern limit of their great line of flight.

The migration of birds in the western hemisphere is a subject of enquiry concerning which, little progress has yet been made, and which for many years to come will require the close observations of naturalists to compass its vast extent. In the pages of this little work, the reader will find a clear and interesting account of the migration of that remarkable wanderer, the golden plover of America, by a gentleman long resident in the Bermudas; and which, we think, will tend to throw considerable light upon the principle of migration. To gain a thorough knowledge of the marvellous movements of the feathered race, intelligent observers should be selected at various distant localities, and as many birds may be supposed to follow the land in preference to crossing the sea, the Isthmus of Darien should on no account be neglected as an important point of observation: thus, by the united efforts of many, the mysterious cloud which at present veils the wonders of migration, may be dispelled, to shew forth in still brighter splendour the stupendous power of Him who ordereth all things so wisely and so well.

THE NATURALIST IN BERMUDA.

BOSS' COVE.

GEOLOGY.

THE geological interests of the Bermudas are peculiar, and by no means inconsiderable, as they belong to the class of "Formations now in progress." They, like the Bahamas, are plainly Coral formations below low-water mark, and Æolian above it. The vital powers which produce the former, and the mechanical action of the wind to which the latter are referable, being still in action. Hence, with two possible exceptions, no traces of the remains of even *locally* extinct animals or plants have yet been discovered, or can reasonably be expected. Some years ago, a very valuable Paper on the Geology of the Bermudas, appeared in the Transactions of the Geological Society of London.

It was written by Lieut. (now Col.) Nelson, of the Royal Engineers; from which, as it has been acknowledged to be the best treatise ever produced on the subject, we cannot do better than glean that portion which is suited to our purpose, in order to shew to what particular causes the Bermudas owe their origin.

This officer informs us that the whole of the islands are composed of a calcareous sandstone, derived from comminuted shells and corals, varying in texture from loose sand to a hard compact limestone; the varieties being irregularly associated, and without any order of super-position. From the saddle-shaped sections observable throughout the islands, he concludes that the strata have evidently been accumulated round numerous and contiguous centres. The bottom of the basin consists of coral reefs (which do not appear above low-water mark, except at spring-tides,) and calcareous sand; the latter being associated with considerable tracts of chalky ground, in which the best anchorages are found. He met with this chalk, likewise, as an occasional constituent of the rock, lying irregularly among its beds in a soft state, or lining caverns, but seldom exceeding a few square yards in surface, and a few inches in depth.* The only minerals noticed were small pieces of oxide of iron, of very questionable origin; menaccanite, found near the Ferry, between St. George's Island, and Bermuda or Main Island; arragonite (?); and a minute quantity of manganese in the red earth.

* The Royal Engineers claim to be the first to give any distinct clue to the origin of chalk; though within but a few months they were followed by Darwin and Lonsdale, on other and perfectly independent grounds. See Bahama Paper, Geol. Soc. Transactions. By Colonel Nelson.

We now quote at length from Colonel Nelson's paper:—

"GEOLOGICAL DETAILS OF THE PRINCIPAL ISLANDS, IN GEOGRAPHICAL ORDER.

"ST. GEORGE'S ISLAND.—The strata dip southward with considerable uniformity along the southern and eastern shores, which are protected from violent attacks of the sea by St. David's, Smith's, and Long-bird Islands; and by the reefs which bound the intricate passage into Murray's Anchorage.

"The north side is exposed throughout its whole length to the fury of a sea which has had, within the area of the north reef (at a distance of seven or eight miles), sufficient space to re-accumulate its destructive energies; and exhibits in the cliffs a saddle or dome-shaped structure, the lines of lamination dipping towards every point of the compass. In one or two instances, the summits of what were once internal hills, are bared; and within a few square yards, the coats of this nucleus range round it with perfect regularity. As might be expected, this north side consists, in a great measure, of abrupt cliffs, and landslips brought down by the undermining of the waves. The sandstone on the summit of some of the hills is scarcely tenacious enough to hold together; but from Mullet Bay to the Ferry, on a level of not more than twenty feet above the sea, the rock suddenly becomes a very hard, fine-grained, or compact limestone, in which scarcely a vestige of organic structure is visible, unless the stone be polished.

"Just above high-water mark, along a considerable portion of the south side, is a stratum of calcareous sand, about six feet thick, apparently a distinct deposit from the rock above it.

"Vertical crevices abound, filled with a substalagmite, much harder than the rock. In some places, the arrangement resembles that of veins in old slate or limestone. Where the partitions stand comparatively unimpaired by the weather, their planes constantly intersect one another, and thus form reticulations, once filled with the soft sandstone. In all cases, the veins are younger than the crevices, from whatever cause these last may have originated.

"ST. DAVID'S.—Like St. George's, the lower level of this island consists of a low, irregular belt of hard limestone, which commences in the centre of the north side, and continues westward through Stock's Point, till it appears, as above noticed, at Mullet Bay.

"COOPER'S ISLAND.—Here I have remarked nothing, except the supposed cast of a tree at the north-east point.

"LONG BIRD ISLAND affords an excellent horizontal section along the south side, of strata deposited in irregular undulations, and not contorted by any subsequent disturbance.

"NONSUCH, CASTLE, and GOAT ISLANDS.—As far as interest depends on variety, these possess but little. They, however, exhibit the frustra of the saddle-shaped structure, which may often be traced from one island to another, though frequently separated by considerable intervals.

"BERMUDA, or the Main Island.—The description given of the north side of St. George's, answers in all respects to that of the whole of those coasts which are exposed to the ocean. On the south side, the sand has made several encroachments, but only at Elbow Bay with any great success. The proprietor of the principal part of the land of this bay, remembers an attempt about seventy years ago, when the inhabitants expected an attack from the French and Spaniards, to form a breastwork along the sand hills, which then, as at Shelly Bay, skirted the coast. In doing so, they cut through the natural protection given by the sea-shrubs and creepers, which usually abound in such places. From that day the sand, supported by constant supplies from the sea, has steadily proceeded up the hill to the very summit, a height of about 180 feet.* There is another encroachment at Tucker's Town, said to have taken place about sixty years ago, and has crossed the neck between Harrington Sound and the sea; but beyond this it does not seem inclined to move. The sand has not been stopped, at the eastern extremity of this beach, where the bluffs commence by their very considerable declivity; though it has been most effectually, at the crest of the slope, by a natural fence of sage bush (*Lantana salvifolia*), growing partly in the soil and partly in the sand; which, as it ascended, seems to have then rolled on with the seeds of this plant, and of destruction to its progress, in its own bosom.

"The same operations appear to have occurred throughout the sand tracts at and near Great Turtle Bay."

Colonel Nelson says the whole of the Bermudas, (and, perhaps, many of the older rocks,) may be called 'Organic Formations,' as they present but one mass of animal remains, in various stages of comminution and disintegration. From the most compact rock to the very sand of the shore, the materials of this group being universally fragments of shells, corals, &c. The *Turbo pica* and *Venus Pensylvanica*, are found imbedded in the rock in great quantities. We

---

* It has long since crossed the brow of the hill.

have specimens of Turbo pica now in our possession, which we obtained on the hilly ground of Smith's parish, adjacent to Hungry Bay; and it puzzled us not a little as to how these large shells could get into that position, being some thirty feet above the level of the sea; but Col. Nelson throws considerable light on the matter, by stating that the soldier-crab is the means of bringing them into these high positions.*

Extensive engineering operations, road-making, deepening channels, &c., have afforded an opportunity never hitherto possessed by writers on coral-belts, throwing much light upon the subterranean formation of the islands; and caverns of various dimensions have been discovered at different times during the progress of the works.

The Colonel states that, whilst cutting the escarp of the North Bastion at Ireland Island, a cavern was opened which contained great numbers of a large and delicate *Helix*, and a heap of the dry red earth of the islands, since determined by him to be mainly of animal origin. In this earthy matter were quantities of birds' bones; and on another occasion, whilst excavating a ditch, a small hole was discovered in a rather hard rock, composed of comminuted fragments, with the interstices not filled up. This hole was about twenty feet above the sea, thirty yards from it, and fifteen feet from the top of the hill, and without any apparent connection with the surface. In this hole were found an egg shell, and many fragments of bones, similar to the preceding; but they were all, as well as the egg, coated with

* This animal has been seen actually climbing up a *door-post*, laying hold of the arris-edge by its claws.

drusy crystals of carbonate of lime. The bird is supposed to have been the "boatswain" (*Phaëton œthereus,*) which resorts to the islands during the breeding season, and makes its nest in the hollows of the cliffs; the bird in this case being immured, during the period of incubation, by some fresh deposit of sand, which had afterwards hardened into the walls of its prison-house. Cases of bones and eggs being found, have occurred in other parts of the islands, imbedded in the limestone: and what may be considered more singular, a gold knee buckle was said to have been found not far from the Tanks on the Main Island; and a canister shot was dug up in coarse limestone, whilst excavating the foundations for one part of the North Bastion, across the bottom of the cavern. Turtle bones have also been found in the loose sand of the sea beach; the turtles sharing the same fate as the bird before mentioned, being buried whilst depositing their eggs. Colonel Nelson was informed by an eye-witness, that the dimensions of the skeletons of these animals were nine feet in length and seven in breadth. He terminates his list of animal organics by stating, that almost every shell known in the surrounding sea may be found in the rock quite perfect.

As regards the Bermuda chalk deposit, Colonel Nelson attributes its existence in this locality to the decomposition of zoophytes, from the least calcareous to the large and massive *Meandrina* and *Astrea.* He states that these animals, and the many marine plants, consisting chiefly of lime, produce this chalk, just as terrestrial animals and vegetables give carbon to soil when they decay.

In various parts of the islands, rudely shaped cylindrical blocks are to be found on the surface; and of these well known curiosities the Colonel says: "They are very frequently arranged in groups of from twenty to fifty, as if they had once been the roots of as many trees. It is the general surmise that these curious cylinders are the remains of the Palmetto (*Chamærops Palmetto, Titf*). I think it very likely that they belong to a member of the family of Palms; though I have my doubts as to the Palmetto being the individual."

To the caverns of Bermuda, which are so remarkable for their singularity and beauty, it will be well to devote a short space; for we doubt if in interest and varied appearance, anything else on the Islands can be compared to them. It would be difficult to describe them, as any account must necessarily fall far short of the reality; but if the reader can imagine an opening of tolerably large dimensions in the limestone rock, and charmingly irregular in outline, from the roof of which shining stalactites descend, reflecting their protracted forms in the light blue-green waters below, which cover the floor of the cavern, and in whose pellucid depths may be seen floating the forms of fishes, garbed in coatings of the most resplendent hues, he will have some idea, albeit a faint one, of the interesting features of these subterranean recesses.

In Mr. Wood's demesne at Walsingham, several examples of the kind may be seen, and the manner in which the hospitable owner of this pretty estate has aided the charms of nature, by means of art, is singularly manifest; the irregular masses of limestone rock, which hang above and around these natural caverns, having various kinds of shrubs, trees, and plants, growing from every crevice. The caverns, however, and the objects around them, are seen to the best advantage, when lit up by the refulgent rays of a 'Mudian' sun; and when the intense heats of summer prove too great for the wearied frame, it is delightful, indeed, to pay a visit to these cool, calm scenes of Nature.

We were informed by Mr. Wood, that he believed his property to be perfectly undermined by caverns, and, doubtless, there are many in various parts of the islands which remain for the investigation of future years. Col. Nelson, in his interesting paper, says, the largest cavern now known, and geologically speaking, the most instructive, is Bassett's, near Somerset Bridge, which is said to extend for more than a mile. He says it is comparatively recent, from the fresh state of its surfaces, and the small quantity of stalactite observable; he accounts for the origin of this cavern by the undermining of the substrata by the sea, the waters of which lie in pools at the bottom; "hence," says the Colonel, "we may consider such caverns as hollows produced by internal landslips, from the most normal of which, to the simplest niche, there is every intermediate point of transition." The most exquisite bijou of all the Bermuda Caverns is that in Tucker's Island; but it must be seen by several well placed candles, from the little boat, which must be launched into this beautiful cave.

"To the unequal distribution of carbonate of lime in solution, which forms ordinary rock on the upper parts of a section, or druses and breccias, as it filters through the strata of red earth to the lower beds, or to the caverns, where it crystallizes as stalagmite and stalactite," the Colonel attributes not only the caverns and sandflows, but the pinnacled rocks, almost equally common in the islands. At Tobacco Bay, (St. George's,) is a most curious group.

The North Rock, which lies some nine miles to the northward of the islands, is obviously the last remnant of a former island; the tops of the pinnacles are stratified. In old French charts, the North Rock is called "Les Petites Bermudas." According to Williams, in his History of the Bermudas, this rock, or rather rocks,—for there are four needle-like structures,—stand ten feet above high water, and vary from four to about eight feet in diameter, and, when struck by the hand, they ring, being sonorous in a remarkable degree.

SPANISH POINT.

ZOOLOGY.

MAMMALIA.

THE Mammals of Bermuda are few in number as regards species, and, with the exception of domestic animals, may be comprehended in the following list, viz.:—*Mus musculus*, Common Mouse; *Mus decumanus*, Common Brown, or Norway Rat; *Mus rattus*, Black Rat; *Vespertilio pruinosus*, Hoary Bat; *V. noctivagans*, Silver-haired Bat; *Balæna mysticetus*, Common, or True Whale; *Physeter macrocephalus*, Sperm, or Spermaceti Whale.

The rats and mice have evidently found their way to the Islands in ships; and once gaining a footing, as all well know, their increase is only a natural result. The former exist in great numbers in some parts of the islands, where they are especially destructive in the sweet potatoe fields.

"Mus decumanus," says Mr. Hurdis, "swims and dives remarkably well, eats fish, and almost anything else it comes in contact with. My little boy was present with a fishing party, when a large one was speared in the water. I once witnessed confusion in their camp, arising from an unusually high spring tide flooding the main ditch near the Speaker's residence. The rats, driven from their holes by the water, were running about like wild rabbits, and I bagged no less than five large fellows, as food for a captive peregrine falcon in my possesion. They sometimes attain large dimensions, and Mr. Fozard once sent me the largest specimen of the Norway rat I ever saw." May, in his account of the shipwreck of Sir George Somers, states, that at that date, A. D. 1609, no rats or mice were to be found in the Bermudas.

Bats may be considered as rarities, and are only observed at a particular season of the year. We cannot do better than introduce here a few notes on this animal, by our friend Mr. Hurdis: "Having for several years noted in my journal the appearance of bats in the Bermudas, and being convinced that the visits of those animals are periodical, I record the following observations with a view of shewing upon what grounds this opinion has been formed.

"Bats are unknown in the Bermudas during the greater portion of the year, and have never been known to breed there. I have noted a solitary instance of one being captured at the close of August, in the year 1849; the middle of September, however, is the season at which they usually appear; from that period to the end of December, these curious animals may occasionally be observed, at evening twilight, in vigorous pursuit of their insect prey, turning and twisting in a most extraordinary manner while so engaged, and affording ample proof of the wonderful power of wing with which Nature has endowed them. On the 30th of January, 1852, a bat was observed near the town of Hamilton, and from being the only instance of its occurrence in that month, was probably a straggler.

"From the circumstance of the only two species yet obtained being common to North America, viz.,—the Hoary Bat (*Vespertilio pruinosus*), and the Silver-haired Bat (*V. noctivagans*), it may be considered beyond a doubt that these interesting strangers come from that portion of the globe.

"Admitting such to be the case, let us enquire how these aeronauts have been enabled to cross an expanse of six or eight hundred miles of ocean in order to reach the Bermudas; and whether we should regard them in the light of accidental travellers blown off the American coast, or as creatures guided by the hand of an Allwise Providence, seeking instinctively, as we call it, a more southern abode.

"The following entry occurs in my journal, Dec. 14th, 1851.—On my way home I was joined by my neighbour Mr. B——, and the master of an American vessel, recently arrived in distress from loss of sails. On allusion being made to the bats I had observed on the road, the American skipper mentioned, that in the preceding month of October, while on a voyage from Newport to South Carolina, his vessel being from 150 to 200 miles from the American coast, and the breeze blowing strong from the westward, a bat came on board his vessel during the night, and was captured by the seamen. He described it as being of a reddish-brown colour.

"Here, then, is an instance of the bat being found in the act of traversing the ocean; but whether blown off the coast, or migrating to more southern latitudes, is a problem which can only be solved by further observations, and an improved acquaintance with the history of this remarkable animal. May we not infer that its route, thus far, had been performed subsequently to the sunset of the previous evening, and that if the animal's course had not been arrested by falling in with the "Warren Brown," it might have continued its flight to the lonely islands of Bermuda, or even to far more distant lands?

"That the bat genus does cross the ocean from the shores of America to the Bermudas, I regard as an established fact, proved by the periodical visitations I have already alluded to; and if further proof of its power of flight should be required, I would refer to the well-authenticated circumstance of a specimen of *V. pruinosus* having been captured in South Ronaldsha, one of the Orkney Islands, in 1847, as mentioned in the Zoologist of January, 1849, there being no reason to doubt the actual flight of this animal across the Atlantic in the month of September.

"Now, if an animal so local in its habits as the bat is supposed to be, should be accidentally blown off the American coast, and compelled to wander over the ocean until it reached the Bermudas, one might be led to suppose that the mild climate of those islands, with the abundant store of insect food at command, would be so completely congenial with its nature, as to induce the stranger to become a permanent resident in its new abode, and relinquish altogether the desire of repeating what to many may appear a dangerous and fatiguing flight over the waters of the Atlantic. Such, however, is not the case; the bat visitors disappearing, as already stated, about the end of December. Whither they go I am not prepared to say, though I am inclined to believe that they continue their course to the southward. This simple fact appears to me to set at rest the supposition of bats being accidentally blown off the American coast. Is not the cause or impulse which dictates this departure from the shores of Bermuda, of the same mysterious character as that which influences the periodical migration of the feathered tribes?

"I have spoken of the autumnal appearance of the bat, because at that season of the year only is it generally to be met with in Bermuda. On two occasions only have I observed a deviation from this rule, the first on the 23rd of April, 1849, when two of these creatures appeared, busy on the wing over a secluded pond in Paget's parish, one of which I shot; the second on the 17th of March, 1852, when a solitary *V. pruinosus* was met with in the Pembroke marshes. As these observations were made in the spring, when many of the feathered tribe are moving on their northern flight, and when, in the former case, several species had actually reached the islands, it became a problem in my own mind, whether the bats might not also be travelling in the same direction.

"The bat is by no means a common animal in the Bermudas. In some years it is rarely seen, in other seasons it is more common. I have met with several of the native inhabitants who had never seen one.

"It may be inferred from the foregoing observations that *V. pruinosus*, and perhaps one or two other species of North American bat, may be found to inhabit the West Indian Archipelago, and the northern coast of South America, a point I had not the means of ascertaining while in the Bermudas. Should these observations and surmises, however, be confirmed by future research, I shall claim to myself the discovery of the roving habits of the genus *Vespertilio* of the western hemisphere.

"The Silver-haired Bat, (*V. noctivagans*,) which was captured alive near Hamilton on the 8th of Oct., 1850, measured eleven inches in extent, by three and a-half in length, including the tail; the tip of the tail extending beyond the membrane. The body was covered with long hair of a black, or very dark brown hue, tipped here and there with white, particularly from each side of the neck to the posterior portion of the back, forming a whitish band in the form of the letter V; the under surface was of shorter fur, sprinkled all over with white hairs.

"This specimen, which is now in my possession, is believed to be the only instance of this species being captured in Bermuda."

It appears by the account of May, who published a small work detailing the shipwreck of the "Sea Adventure," which took place in July 1609, (a copy of which book may now be seen in the library of the British Museum,) that Hogs were then found in a wild state upon the Islands in great abundance, Sir George Somers having killed thirty-two in one day's hunt.

CETACEA.—The seas around the Bermudas have been known from the earliest period of their history as very productive in this valuable mammal; and at the present day the Americans, if not the 'Mudians' themselves, reap no small benefit from the fishery.

Whale beef is eagerly sought after by the coloured population of the Bermudas, and the cutting up of a carcase, after the operation of *flinching*, is a scene that few would desire to witness a second time.

In towing a dead whale into port, the carcase is not unfrequently followed by its young calf, which never fails to fall a sacrifice to epicurean taste; the beef, or rather veal of the young animal, being considered good eating.

A gun, of peculiar build, is sometimes used in the destruction of these animals. It is a short weapon, with a thick heavy barrel, and is fired from the shoulder, charged with a hollow and pointed metal tube, which explodes inside the whale, causing its instantaneous death. Instances have been known of the tube passing completely through the unhappy creature.

The TRUE OF GREENLAND WHALE (*Balæna mysticetus*) is common on the east and south shores of Bermuda, where it is taken with boats, from March to July.

The SPERMACETI WHALE (*Physeter macrocephalus*) seldom frequents the waters near the shore. One of these whales was captured off St. David's Head, on the 28th of July, 1851; a rare occurrence in the Bermudas, though the American whale vessels annually kill some of these valuable creatures within sight of the Islands.

In a communication to the Royal Society of London, written by a Mr. Richard Norwood, from Bermuda, and bearing date June 18th, 1667, we have the following concerning the whale tribe.—"The killing of whales, it hath been formerly attempted in vain, but within these two or three years, in the spring time and fair weather, they take sometimes one, two, or three in a day. They are less, I hear, than those in Greenland, but more quick and lively; so that if they be struck in deep water, they presently make into the deep with such violence, that the boat is in danger of being haled down after them, if they cut not the rope in time. Therefore they usually strike them in shoal water. They have very good boats for that purpose, manned with six oars, such as they can row forwards or backwards as occasion requireth. They row up gently to the whale, and so he will scarcely shun them, and when the harpineer, standing ready fitted, sees his opportunity, he strikes his harping iron into the whale, about or before the fins, rather than towards the tayl. Now the harping irons are like those which are usual in England in striking porpoises, but singular good mettal, that will not break, but wind, as they say, about a man's hand. To the harping iron is made fast a strong lythe rope, and into the socket of that iron is put a staff, which, when the whale is struck, comes out of the socket, and so when the whale is something quiet, they hale up to him by the rope, and, it may be, strike into him another harping iron, or lance him with lances in staves till they have killed him. This I write by relation, for I have not seen any killed myself. I hear not that they have found any spermaceti in any of these whales, but I have heard from credible persons that there is a kind of such as have the *sperma*, at Eluetheria and others of the Bahama Islands, (where also they find often quantities of ambergrease,) and that those have great teeth (which ours have not) and are very sinewy. One of this place, (John Perinchief,) found one there dead, driven upon an island, and though I think ignorant in the business, yet got a great quantity of spermaceti out of it. It seems they have not much oyl, as ours; but this oyl, I hear, is at first all over their bodies, like spermaceti; but they clarify it I think by the fire. When I speak with him (whom I could not meet with at present, and now the ship is ready to sail), I shall endeavour to be further informed; but at present, with the tender of my humble service to the Royal Society, and commending your noble designe to the blessinge of the Almighty, I take my leave," &c.

Another communication to the Royal Society, from a Mr. Richard Stafford, and dated from Bermuda, July 16, 1688, has the following:—"We have hereabouts very many sorts of fishes. There is amongst them great store of whales, which in March, April, and May, use our coast. I have myself killed many of them. Their females have abundance of milk, which their young ones suck out of the teats, that grow by their navel. They have no teeth, but feed on moss growing on the rocks at the bottom during these three months, and at no other season of the year. When that is consumed and gone, the whales go away also. These we kill for their oil. But here have been spermaceti whales driven upon the shore, which sperma (as they call it), lies all over the body of those whales. These have divers teeth, which may be about as big as a man's wrist, and I hope by the next opportunity to send you one of them. I have been at the Bahama Islands, and there have been found of this same sort of whales dead on the shore, with sperma all over their bodies. Myself, with about twenty more, have agreed to try whether we can master and kill them; for I could never hear of any of that sort that were killed by any man, such is their fierceness and swiftness. One such whale would be worth many hundred pounds. They are very strong, and inlayed with sinews which may be drawn out thirty fathom long."

Perhaps it may be interesting to some of our readers, if we insert here the particulars of a "whale case," which was tried at the Bermuda assizes, in the autumn of 1857. It was obligingly communicated by a gentleman, who has kindly assisted us with information for this little volume:—
"Each company of whalers has two or more boats. One of these companies had the good luck during the last season to fall in with a valuable Cape whale, which yielded about £200 worth of oil. The boat which first struck it was split; her companion was soon at hand, and struck into it also, cutting away, of course, the first tine; that boat, also, was injured, and, by some singular chance, the whale, with the boat fast to it, found its way into Castle Harbour. Either from fear or mis-management, the crew did not pull up to the fish to lance it, the whale running about to and fro, and (as they term it), 'feeling for the boat.' After being fast an hour, or longer, a small boat put off from the shore, belonging to another Company; the harpooner pulled a lance from the 'fast' boat, and went alongside the finny monster, and plunged the lance once or twice right into a vital part, and killed the whale. He claimed allowance for 'work and labour,' which was inhumanly denied him; a law-suit was the consequence, and the jury awarded him fifty pounds."

The domestic animals of the Bermudas are cattle, horses, mules, donkeys, sheep, goats, pigs, dogs, rabbits, and cats.

The native cattle consists chiefly of cows and heifers, many of which, from a curious formation of the horns, appear to be descended from some foreign breed. They are of small size, and manage to subsist where a larger and heavier animal would starve. Fresh butter, of excellent quality, is made from these animals, but not in any quantity. A few draught oxen, imported from the United States, are also to be met with, in ploughs and bullock carts. Bullocks, for the supply of the Military, Naval, and Convict Establishments, are also imported from the United States, as well as from the British North American Colonies, at the rate of about one hundred head per month. These are stall-fed by the contractor till slaughtered, their food being imported hay, and Indian corn meal.

The local government has, of late, endeavoured to improve the breed of cattle, and has gone to considerable expense in procuring bulls and cows from Alderney; but we have yet to learn whether they thrive sufficiently well to make amends for the outlay.

Horses are obtained chiefly from the United States, and make very useful hacks; and, although it cannot be said that they are remarkable for sleekness or symmetry, still they suffice for the requirements of a colony, which has the good sense to study the useful in preference to the ornamental. Handsome horses, nevertheless, are to be obtained from the States, if handsome prices be given for them.

Mules, of a superior kind, are occasionally imported from the United States.

Donkeys, with carts and harness complete, are not unfrequently imported from England by masters of Bermudian vessels, who realise a handsome profit by the speculation.

These animals are extremely well adapted to the climate and scanty fare of the Islands, and, from their very great utility, should be more generally encouraged.

Sheep, of a very ordinary description, are also imported from the United States, and, sometimes, from the North American Colonies, to supply the market. They are seldom seen beyond the limits of the importers' sheep pen.

Goats are kept by most housekeepers, who do not allow them to roam at large, but usually tether them, with some yards of rope, to a stake driven in the ground, on some grassy patch, where they nibble away through the day, and at night are taken up and secured in an outhouse. They are generally very tame, and are frequently made pets of, eating from the hand almost everything offered. They are also extremely susceptible of cold; and when a cool, rainy day in autumn sets in, they shiver all over, and bleat continually, until removed under shelter. The islands are well adapted for these animals, and they thrive on the stunted herbage, which an English farmer would look upon with a melancholy eye.

Pigs are imported from the United States, and likewise reared in the islands. When fatted, (with the aid of American Indian corn and meal,) this native pork sells at a lower price than ordinary butchers' meat.

Regarding the rabbit (*Lepus cuniculus*) Mr. Hurdis sends us the following note:—

"On the 22nd January, 1851, rowed to one of the uninhabited islands of the Great Sound, in company with Mr. Hodgson Smith, for the express purpose of 'rabbit shooting.' Found several of those animals among the sharp rocks and densely growing stunted cedars, and succeeded in killing four couple. They were all of the domestic kind,—of a yellow, grey, and black colour; and, although the herbage was exceedingly scanty, of good size, and excellent condition.

"When the scanty supply of grass is burnt up by the heat of summer, I am inclined to think that these rabbits subsist by barking the under side of such branches of cedar (*Juniperus Bermudiana*), as grow within their reach; at all events, the state of the cedars induced that belief.

"This island, like its neighbours in the Sound, is entirely without water."

AVES.

NOTES AND OBSERVATIONS ON THE RESIDENT AND MIGRATORY BIRDS OF THE BERMUDAS.

BY MAJOR J. W. WEDDERBURN (late 42nd Royal Highlanders).

TURKEY BUZZARD (*Cathartes aura*). My friend, Mr. Hurdis, examined a male specimen of this bird, on the 29th December, 1853. It was shot by Mr. George Trimmingham, in the latter part of November of that year, at

Mr. Harry Tucker's ponds, and presented to Dr. Monroe (Staff Surgeon), then in charge of the sick, at Prospect Hill encampment.*

OSPREY (*Pandion haliœtus*). This bird is said formerly to have been abundant in the Bermudas; of late years, however, it had not been noticed till October 15th, 1847, when a very fine specimen was shot by Colonel Dunsmure (late 42nd Royal Highlanders), at the Sluice Ponds, which is now in my possession. This species was frequently seen afterwards, along the shores of the different islands, and several specimens were obtained. The Sluice Ponds, which abound with grey mullet, were one of their most favourite haunts. On one occasion, when out there with my gun and casting net, I threw the latter over a very fine osprey; the bird was so wet after his day's fishing, that he could hardly fly; he managed, however, to escape from the net, but fell to my gun.

PEREGRINE FALCON (*Falco peregrinus.*) The first notice of this bird was in 1846. It was killed by Dr. Cole (20th Regt). I often observed one flying about in January, 1850, and on the 1st of February following, it was wounded by a 'Mudian,' and given to Mr. Hurdis, who kept it alive for a long time; its ultimate fate I have forgotten.

PIGEON HAWK (*Falco columbarius.*) This pretty little falco is found occasionally, throughout the year, on all the islands; but it has not been known to breed.

WILSON'S SPARROW HAWK (*Falco sparverius.*) A beau-

* The head of this bird is now in the collection of our friend, Capt. Edward Loftus Bland, of the Royal Engineers, to whom that portion was presented by Dr. Monroe; the greater portion of the skin, not being well preserved, was finally thrown away, and the above part only kept.

tiful specimen of this hawk, was killed near the Sluice Ponds, 9th December, 1853, whilst in the act of pouncing on some chickens.

SHARP-SHINNED HAWK (*Astur fuscus.*) A single specimen, shot near Peniston's Ponds, on the 23rd February, 1853, and kindly sent home to me by Mr. Hurdis, and now in my collection.

THE HARRIER (*Circus cyaneus.*) Occasionally seen during the autumn migration, although I never had the good fortune to meet with it in any of my constant rambles over the islands.

SHORT-EARED OWL (*Otus brachyotus*). One specimen shot by Dr. Cole (20th Regt.), in December, 1846.

LONG-EARED OWL (*Otus vulgaris*). Mr. Hurdis mentions three of these birds as having been killed at Gibbs' Hill, in 1846, 1847, and 1849; the one obtained in the latter year, is now in my possession.

SNOWY OWL (*Surnia nyctea*). Lieut. Fayrer, R.N., shot two specimens at Boss' Cove, in the autumn of 1843. Another, a fine female specimen, was shot by a person named "Llewellyn," at Ireland Island, on the 29th of November, 1853; this bird was only wounded, and when examined by Mr. Hurdis, on the 13th December following, it appeared lively and well. When being fed, it frequently erected a little tuft of feathers on each side of the head, so as to resemble small horns.

ACADIAN NIGHT OWL (*Ulula Acadica*). Only one specimen, found January 12th, 1849, sitting inside the muzzle of one of the guns at Ireland Island, by an artillery man. It is to be hoped that the said gunner has more nerve when working a gun, than he displayed on finding the little bird, being afraid to catch it, as he said, "it glow'rd at him." It was caught by a man of the 42nd, and lived in my room for several days, getting quite tame. At night it always became restless, and finally killed itself against the wires of its cage. Mr. Harry Tucker saw another some short time afterwards, in a cave on the south shore.

GREAT NORTH AMERICAN SHRIKE (*Lanius borealis*). Only one specimen shot by myself, on the 12th of March, 1850, near Harris's Bay, and now in my collection.

HOODED FLY-CATCHER (*Myiodioctes mitratus*). One specimen only, shot at Ireland Island, March 30th, 1847.

PIPIRY FLY-CATCHER (*Muscicapa dominicensis*). The first specimen was found by myself sitting on the top of a cedar tree, in Mr. Hurdis' garden, on March 30th, 1850. It proved to be a very fine male specimen. Two others were killed on St. David's Island, 15th of April, of the same year.

TYRANT FLY-CATCHER (*M. tyrannus*). Very numerous in April, 1850. They were found in all the swamps, and were extremely tame. I generally saw them sitting on the fences, and darting after passing insects, always returning to their former station.

WOOD PEWEE FLY-CATCHER (*M. vireus*). A single specimen, shot by Mr. Hurdis, near Chief Justice Butterfield's ponds, on the 30th April, 1852.

WOOD THRUSH (*Turdus mustelinus*). Found by Colonel H. M. Drummond (late 42nd Royal Highlanders), near St. George's, in 1849, and now in my collection. Several others were afterwards obtained.

OLIVE-BACKED THRUSH (*T. olivaceus*). Two specimens only; one, by Colonel Drummond; the other, by Captain McLeod, of the same Regiment, in 1849.

Migratory Thrush (*T. migratorius*). On the 26th of February, 1850, I found a small flock of these birds, amongst the cedar trees, in Pembroke Marsh, and succeeded in killing three of them. A few others were afterwards shot; the remainder were spared, as I hoped they might have remained to breed; but they all departed on their northern migration a few weeks afterwards, not leaving even a straggler.

Cat Bird (*Orpheus Carolinensis*). The mocking bird of Bermuda, is very common all the year round, and breeds. Its nest is composed generally of dry weeds, and small twigs, and lined with fibrous roots. They lay from four to five eggs, of a bluish green colour.

Aquatic wood Wagtail, or Water Thrush (*Seiurus noveboracensis*). This little bird is so extremely shy, that, although not uncommon, it may yet be far more numerous than is supposed. Its sharp, yet sweet little cry, may be heard in all the mangrove swamps, particularly at Riddle's Bay, the Sluice Ponds, Hungry Bay, and in Somerset Island. If one attempt to approach near it, it darts off so quickly and quietly amongst the roots of the mangroves, that it is almost impossible to get a shot. I, however, obtained several specimens of it. Several times, at Riddle's Bay, I have noticed, seemingly, a larger species of this bird, but never succeeded in killing any of them. They appear regularly in autumn, and a few remain throughout the winter.

Baltimore Oriole (*Icterus Baltimore*). Capt. Tolcher (56th Regiment), shot one of these birds early in October, 1854, at Somerset. Mr. Hurdis found it amongst his collection of skins, on the 20th of same month, when

Captain Tolcher assured him that Mr. Harford, of his Regiment, had killed another specimen about the same time, which, from being very much mutilated, he had unfortunately thrown away.

BLUE BIRD (*Sialia Wilsonii*). Very common all the year round. They sometimes appear in much greater numbers, in various parts of the islands, especially in January and February. This may partly be from birds migrating from the American coast. On January 5th, 1848, I noticed them in large flocks, at Ireland Island. I shot a nearly white variety, on the 30th May, in the same year.

WHEATEAR (*Saxicola œnanthe*). Mr. Hurdis mentions one as having been shot by Lieutenant Wood (20th Regiment), near the lighthouse, on the 5th October, 1846; unfortunately, only the tail feathers were preserved; these were, however, sent to the late Mr. Yarrell, who said there could be no doubt as to the identity of the bird. Another, a female, was frequently seen by Colonel Drummond and myself, on the north shore, near St. George's, in March, 1850; but it was so very shy, we could not manage to shoot it.

YELLOW-CROWNED WOOD WARBLER (*Sylvicola coronata*). A few specimens have been obtained. I killed two in Pembroke Marsh; and Mr. Hurdis shot one at Hungry Bay, January 24th, 1850.

PINE-CREEPING WOOD WARBLER (*S. pinus*). First noticed, September 27th, 1849, when a good many appeared on the Islands; they all disappeared by the beginning of October. Several of these birds were captured outside the lantern of the lighthouse, in the dark and rainy night of the 5th of September, 1850.

YELLOW-RED-POLL WOOD WARBLER (*S. petechia*). Only

two specimens, shot by myself on 17th December, 1847, and 3rd of December, 1848, in Pembroke Marsh.

BLUE-YELLOW-BACKED WOOD WARBLER (*S. Americana*). The Rev. H. B. Tristram killed the only specimen hitherto obtained, on the 21st April, 1849, at Ireland Island.

PRARIE WOOD WARBLER (*S. discolor*). Shot by myself in the dockyard, at Ireland Island, whilst on guard, 3rd October, 1848, contrary to all the rules of H. M. S., and of Mr. Ballingall's. This little bird I had noticed for some days previously, but could not get near it, it was so shy. This is the only specimen hitherto obtained.

BLACK-AND-WHITE CREEPING WARBLER (*Mniotilta varia*). One example only of this pretty bird, shot by Colonel Drummond, at St. George's, in October, 1849, and now in my possession.

MARYLAND YELLOW-THROAT (*Trichas Marylandica*). A male specimen was shot by Mr. Hurdis, in a bushy swamp, near the Sluice Gates, on the 18th October, 1853. It was, unfortunately, very much mutilated.

CEDAR WAXWING (*Bombycilla Americana*). I first found this bird on the 10th October, 1847, at the Ponds, near Hungry Bay; there were about thirty of them, flying about amongst the cedar trees, three of which I shot. On the 17th December following, I shot four others, one of which had a few of the beautiful wax-like tips to the secondary feathers. In December, 1849, Mr. Hurdis killed two out of a flock of about twelve. On the 5th January, 1850, I noticed one in the garden, at the officers' mess house, at Hamilton.

AMERICAN PIPIT (*Anthus ludovicianus*). One specimen only, shot by Mr. Fozard, on the 26th November, 1848, and now in my possession.

SHORE LARK (*Alauda alpestris*). The two first specimens of this bird were shot by Colonel Drummond, at St. George's, 25th October, 1849. I observed another for a long time at Spanish Point, and at last succeeded in shooting it, 27th February, 1850.

EUROPEAN SKY LARK (*A. arvensis*). Mr Hurdis shot a beautiful specimen, on the north shore of Hamilton, on the 12th of June, 1850. This bird may have escaped from an emigrant ship, but having no marks of being a caged bird, and both the British landrail and wheatear having been found, it may safely be added to the Bermuda list. A strong easterly gale must have driven it out of its course. When seen, it was soaring high in the air, and its well known song first attracted Mr. Hurdis' attention.

SNOW BUNTING (*Plectrophanes nivalis*). On the 18th of February, 1848, when sailing past Grace's Island, I noticed a small flock of these birds, and fortunately killed one of them; the wind blowing pretty strong from the north-west, in which quarter it had been for ten days previously. Many others were afterwards killed on the glacis of the fortifications, both at Ireland Island, and St. George's. I noticed four others at Ireland Island, on the 4th January, 1850.

BAY-WINGED BUNTING (*Emberiza graminea*). One example only, shot by Captain McLeod, at St. George's, 25th October, 1849, and now in my possession.

SAVANNAH BUNTING (*E. savannah*). I killed the only specimen of this bird hitherto obtained, on the 11th April, 1850, in Pembroke Marsh.

HENSLOW'S BUNTING (*E. Henslowi*). Mr. Hurdis shot one specimen out of a small flock of these birds, in the Pembroke Marshes, on the 2nd of December, 1850. They had frequented the dense reeds and rushes for a fortnight previously. This specimen is now in the collection of Captain Orde (late 42nd Royal Highlanders).

RICE BUNTING (*Dolichonyx oryzivora*). First noticed by Mr. Hurdis, on the 13th October, 1847, when he killed three out of a small flock in Pembroke Marsh; a few remained till the end of the month. During some years this bird is not uncommon, but always found in winter plumage. They are so extremely fat, that it is almost impossible to preserve them.

SWAMP SPARROW (*Ammodramus palustris*). Captain Mc Leod killed one specimen of this bird, in Pembroke Marsh, 3rd December, 1849; now in my possession.

CARDINAL GROSBEAK, or RED BIRD (*Pitylus cardinalis*). Very common, breeds, and remains all the year. I never, during the whole time I was in Bermuda, heard them sing in any note, but one resembling *tew, tew, tew*, repeated three times; the bird generally perched on the top of some cedar tree; nevertheless, he is called the "Virginian Nightingale," and much praised for the clearness and sweetness of his song.

ROSE-BREASTED GROSBEAK (*Coccoborus ludovicianus*). A female specimen was shot by Colonel Drummond, on the 9th October, 1849, near St. George's. On the 15th April, 1850, when out near the residence of W. B. Smith, Esq., (the Treasurer,) Mr. Hurdis shot a magnificent male specimen. Both are now in my collection.

SUMMER RED BIRD (*Pyranga æstiva*). A female specimen was shot on the 9th April, 1850, and on the 19th of the same month I shot a beautiful male specimen; a few others were got during the month, two by myself at Harris's Bay, on the 20th April.

SCARLET TANAGER (*P. rubra*). Two or three examples

were killed in April, 1850; they seemed to have arrived at the same time with the summer red birds. Mr Hurdis gave me a male specimen, which had been obtained near the south shore.

LESSER REDPOLE (*Linaria minor*). Dr. Cole (20th Regiment), shot a specimen on the 8th February, 1847; I killed another at the upper part of Hamilton water, 11th October, 1847; and one or two other specimens were afterwards obtained; one, on the 24th January, 1850, by Mr. Marriott, of H. M. Customs.

THE CROSSBILL (*Loxia curvirostra*). A specimen of this bird was captured in the dockyard, at Ireland Island, 20th January, 1850, and got quite tame, and lived for several days in my room, but poisoned itself by eating part of a composite candle, which it had nearly cut in half with its strong bill during the night. I shot three specimens near Mr. Ewing's house, 5th April, 1850, and saw a small flock on several occasions, near Pitt's Bay, but they were so shy, I could not get near them. They disappeared early in May.

WHITE-WINGED CROSSBILL (*S. leucoptera*). Mr. Falconer presented Mr. Hurdis with a beautiful male specimen, which one of his pupils had killed on the 11th May, 1852. This bird is now in my collection; it was in the mottled plumage of carmine and black, with a touch of yellow on the rump. The same lad who killed this bird had another in his possession, in grey plumage. Mr. Hurdis also examined a female specimen, preserved by Lieut. (now Capt.) Clutterbuck (56th Regt.), which had been shot at Somerset, in March, 1852.

AMERICAN CROW (*Corvus Americanus*). A few of these birds are generally to be seen, between the lighthouse and

Hamilton. I have never found their nests, but they are known to have bred, as a few young crows were observed near Warwick Church, during the first week of April, 1849. It is supposed they were introduced from Nova Scotia some few years ago.

YELLOW-BELLIED WOODPECKER (*Picus varius*). In general, not very common. I first saw it in December, 1847; again, in November, 1848; and in April, 1850, a great many suddenly appeared, several of which I shot. Many of the palmetto trees are bored by this bird. It breeds in Mr. Ballinghall's garden every year; and I should think that a few also breed in holes in the large trees at Brackish Pond, and in some of the other large swamps.

YELLOW-BILLED CUCKOO (*Coccyzus Americanus*). First observed by Dr. Cole (20th Regiment), in 1844. On the 16th October, 1847, another specimen was captured at Ireland Island, and sent to me; and again, another in 1848. An extraordinary flight of these birds occurred on the 9th of October, 1849, thousands of them appearing, the most of which disappeared the following day. A few again appeared, about 23rd April, 1852, remaining for two or three days only.

BELTED KINGFISHER (*Alcedo alcyon*). These birds arrive regularly about the middle of September, and are to be found in all the mangrove swamps, creeks, and ponds, in the islands. Many remain during the winter, but they all disappear about the middle of April. A curious circumstance happened to me on the 5th of April, 1850, when out shooting at Hungry Bay. I wounded a belted kingfisher, which fell into the water, and whilst fluttering about, a large crab darted at it from below, dragging it with him under water. As the bird was struggling a good deal, the crab dropped it, and lay immediately below, till it was quite dead, when he again seized it, and bolted off. Being afraid of losing my bird, I at once gave chase, and speared the crab with the ramrod of my gun, notwithstanding which, he still retained hold of his prey. Having lost one of its claws, I did not preserve it, and very much regret not having done so, as the king-fisher is still in my possession.

Barn Swallow (*Hirundo rustica*). Rarely seen in April and May, but sometimes numerous in August and September. I have seen it as early as 1st of August, in the year 1848, at Hamilton; and they were numerous on that day at Hungry Bay, and Riddle's Bay. This species was very numerous in the great flight of swallows, in September, 1849.

Bank Swallow (*H. riparia*). Two specimens were killed, in September, 1846, by Captain Lye (20th Regiment), and a few were seen near Hamilton, on the 8th of August, 1847.

White-bellied Swallow (*H. bicolor*). A great flight of swallows appeared on the 22nd September, 1849, when this species was numerous, although never previously observed.

Purple Martin (*H. purpurea*). These birds, like the *H. bicolor*, were numerous during the great flight of the 22nd September, 1849, and an addition to the ornithology of the Bermudas.

Spine-tailed Swift (*Chœtura pelasgia*). One specimen only, shot on the 13th September, 1849, and now in Captain J. W. P. Orde's collection. On the 24th of the same month, Mr. Hurdis noticed several of this species, stragglers, left behind, after the great flight of the 22nd September had taken its departure.

Ruby-throat Humming Bird (*Trochilus colubris*). There is a tradition that the humming bird visited the islands of Bermuda, in considerable numbers, about thirty years ago; but of late years they have not been noticed till 26th April, 1852, when Mr. Hurdis wrote to me, saying that Mr. John Darrell, (son of the then Attorney-General, now Chief Justice,) of those Islands, had seen a humming bird under the windows of his father's house, where it was busily employed entering the large white bell-shaped flowers of the giant stramonium, its tail only at times being visible. Another was seen about the same time, and within about two miles of the above place, by a Miss Watson, in her brother's garden.

Virginian Night Hawk (*Chordeiles Virginianus*). These curious birds are sometimes very common in April, and also in September and October, on their migration north and south. The marsh below Government House was their great resort, when just as it was getting dusk, they would appear one by one, and soon be skimming about in all directions, uttering every now and then a sharp, whirring sort of cry. They double and rush about in a most wonderful manner, frequently depressing first one side, then the other; although flying close together, they seem to try and keep apart, each having seemingly its own hunting ground. I had, however, better opportunities of noticing these birds in the West Indies than in Bermuda. They must commit great havoc amongst the mosquitoes (hence its name, mosquito hawk), and other small insects. I saw several of them sitting on the guns of the keep, near the Commissioner's house, Ireland Island, on the 10th of October, 1849. On the 5th of February, 1848, Colonel Drummond saw a few near St. George's. On the 21st of April, 1852, a number of these birds appeared in Pembroke Marsh, remaining a couple of days, and leaving a few stragglers behind, which disappeared soon after them. I have seen the night hawk as late as the 14th of May, 1850. Excepting those already mentioned as sitting on the guns, at Ireland Island, I never saw them during the day time; like our European Nightjar, I suppose they sit *along* the branches of trees, looking so like part of the tree itself, that it is difficult to see them.

GROUND DOVE (*Columba passerina*). Very common; breeds, and remains all the year round.

CAROLINA LONG-TAILED DOVE (*Ectopistes Carolinensis.*) This beautiful bird was supposed to have been seen in Somerset, in 1847. A single specimen was shot by Captain Harvey, of Bermuda, on the 20th of March, 1850, and is now in my collection.

VIRGINIAN COLIN, or QUAIL (*Ortyx Virginiana*). This bird was known to breed in Bermuda, about thirty years ago, but it must now be extinct, none having been seen for many years.*

AMERICAN GOLDEN PLOVER (*Charadrius marmoratus*). During some years, large flocks of these birds pass over the islands, in the months of September and October; but, unless in stormy weather, they do not alight in any great numbers. I have seen it as early as the 21st August, 1847, at Ireland Island; again, on the 25th July, 1848, at Hamilton; the latter was a single bird, sitting on the road, close to the

* Richard Darrell, Esq., (of Lincoln's Inn, eldest son of Chief Justice Darrell,) now resident upon the islands, has recently imported several pairs of these birds from the United States, to turn out; and, by the last accounts, they are increasing rapidly, and spreading over the main island.

house (Wood's Buildings), in which I lived, but by the time I got out it was gone. On the 9th March, 1852, one was shot, in beautiful plumage, on the north shore; and this is the only instance of its appearing in spring.

KILDEER PLOVER (*Charadrius vociferus*). Met with occasionally along the north shore of St. George's, and Hamilton, during winter. Mr. Hurdis shot one as early as the 12th November, 1849. I killed specimens on the 23rd February, 1848, and on the 4th March, 1848. One beautiful specimen, shot and preserved by Colonel Drummond, on the 4th February, 1848, is now in my possession.

AMERICAN RING PLOVER (*C. semipalmatus*). Not uncommon, and to be met with in most of the sandy bays and creeks, particularly at Mangrove Bay, Somerset. They are generally in company with the stints and semipalmated sandpipers.

PIPING PLOVER (*C. melodus*). One specimen only, shot by myself, at Mangrove Bay, Somerset, on the 5th September, 1848. It was in company with a flock of ring plover, and until I shot it, I thought it was merely a white variety of that species. There had been a westerly gale for some days previously, and at the same time I killed this bird, I also got a black-bellied plover, semipalmated sandpiper, turnstones, and Carolina crakes.

BLACK-BELLIED PLOVER (*C. helveticus*). Very rare. The only specimen I know of having been obtained, was shot by me, on the 5th of September, 1848, at Mangrove Bay, Somerset. I saw another, on the 19th of the same month, at the same place, but could not get within shot of it.

THE TURNSTONE (*Strepsilas interpres*). This bird, which is found in every part of the world, is also common on the shores of Bermuda during winter. A single specimen was shot on the 3rd of August, 1849.

THE SANDERLING (*Calidris arenaria*). First found by my friend, Mr. Marriott, the 7th of November, 1847, on the sand hills. Mr. Hurdis killed a very fine specimen, a few days afterwards, at the same place. Another, killed by Mr. Marriott, 4th September, 1848. I have killed a few specimens on the sandy shores of Somerset.

GREAT BLUE HERON (*Ardea herodias*). Many of these birds arrive in autumn, and a few remain throughout the year. In 1846, the nest of this bird, containing two eggs, was found amongst the mangrove trees, at Hungry Bay. The Rev. H. B. Tristram kept one of these herons alive in his garden, at the Parsonage, in Ireland Island, which was once seen to seize a ground dove and swallow it entire.

BLUE HERON (*A. cœrulea*). Two specimens were killed, in 1849, one in full plumage, in October; the other, a young bird, in April. When out shooting with Mr. Hurdis, on the 16th of April, 1850, at Brackish Pond, he shot a splendid male specimen, in full plumage, which is now in my possession. Another was killed on the 15th of May, 1852; this specimen was in the transition state of plumage from white to purple.

GREEN HERON (*A. virescens*). During some years, not uncommon I got one specimen, a young bird of the year, from Mr. Hurdis, the 16th October, 1848. They were very abundant in April, 1849, and again in October of the same year. When on the Staff in Barbadoes, I was very fond of watching these birds feeding early in the morning, just below my windows; they would stand perfectly motionless in the water, every now and then darting their sharp bills into some of the small crabs, with which the little stream was filled. They seemed to catch these crabs and prawns all day long; at other times, they would run along the muddy banks of the river, their necks stretched out close to the ground, chasing the crabs and numerous insects, which were trying to escape from them. In Mr. Gosse's account of the Birds of Jamaica, he speaks of the humming-birds chasing the different kinds of herons. I have also noticed this fact, at a small marsh near Worthing, in Barbadoes, where I have seen the beautiful little blue and green-crested humming-bird chasing the "Green Gaulin," (as it is called,) in every direction, uttering a little, shrill, angry cry, the heron seemingly in great distress, till at last he would take refuge from his diminutive persecutor, by diving into some friendly thicket; the humming-bird would then take one or two casts, backwards and forwards, to make sure his enemy was gone, and then dart off to some flower close at hand.

GREAT AMERICAN WHITE EGRET (*Ardea egretta*). Only two specimens, which were killed at Hungry Bay, in 1840; and Mr. Hurdis mentions that none have been obtained since, though several instances are recorded of its having been seen.

SNOWY HERON (*A. candidissima*). I shot two beautiful specimens, male and female, in full plumage, on the 8th and 12th of April, 1850, at the pond, near the tamarind tree; the male bird was only wounded, and swam out a considerable distance into the pond before I could secure it. Several others visited the islands, in September, 1850. These two birds, and the specimen of *ardea cœrulea,* already mentioned are, I think, the most beautiful specimens I collected in the Bermudas.

AMERICAN BITTERN (*A. lentiginosa*). To be met with in all the marshes, from October till December, and occasionally in March. I killed six specimens, in Pembroke Marsh, in October, 1847. During some seasons, they are rare.

THE LEAST BITTERN (*A. exilis*). I shot a beautiful specimen of this bird, on the 23rd October, 1847, in Pembroke Marsh. A short time afterwards, I killed another at the Sluice Ponds, but, unfortunately, lost it amongst the thick roots of the mangrove trees. I saw two near Boss's Cove, on the 15th of December, 1849; and, again, another, near the same place, on the 17th of the month, but did not succeed in getting any of these. On the 15th of March, 1850, I killed one near Boss's Cove. With the exception of this last bird, I could never get them to take wing; they always ran about amongst the roots of the mangroves, where it was most difficult to find them when once lost sight of. Another example was killed by an officer of the 56th Regiment, in the autumn of 1852.

AMERICAN NIGHT HERON, or QUA BIRD (*A. nycticorax*). This species is not uncommon from the latter part of September, and during the winter till March. I found them generally at the Sluice Ponds, Hungry Bay, and amongst the mangrove trees around the ponds, near Harrington Sound. None have as yet been killed in adult plumage.

YELLOW-CROWNED NIGHT HERON (*A. violacea*). Two specimens of this bird were killed, one, in April, 1848; the other, in September, 1849. Mr. Hurdis gave me a very fine specimen, which was killed on the 3rd of April, 1850, and is now in my possession.

HUDSONIAN CURLEW (*Numenius Hudsonicus*). First found by Captain McLeod, on the 26th of September, 1848, and now in the Rev. H. B. Tristram's collection. It arrives early in August, but is so shy of approach, one can rarely get within shot of it.

ESQUIMAUX CURLEW (*N. borealis*). This bird is not so wary as the last, but is still very difficult to approach. On the 25th of September, 1848, I at last succeeded in killing a very fine specimen at Mangrove Bay, Somerset; and at the same time, three golden plover, one sanderling, and four turnstones, which were all feeding close together. I saw one on the 15th of September, 1848, at Spanish Point, besides various others, which I could not, however, get within shot of.

THE WILLET (*Totanus semipalmatus*). Only one example of this bird, which I shot myself, on Pearl Island, the 3rd of July, 1848.

TELL-TALE TATTLER (*T. vociferus*). These birds arrive in the month of August, and are more or less common in some seasons. On the 4th of August, 1848, I observed a good many of them at Chief Justice Butterfield's pond, in company with a great many yellow-shanks tattlers, stints, and semipalmated sandpipers. On the 20th of September, 1848, I noticed a very large flock going off in a south-easterly direction. They are sometimes met with till the 10th of November.

YELLOW-SHANKS TATTLER (*T. flavipes*). These birds arrive regularly about the 1st of August, in each year, being one of the earliest visitors from the north. It remains till the middle or end of September. On the 13th of July, 1847, one of the men of my company, caught one in an exhausted state, on the north shore of Ireland Island, while half a gale of wind was blowing from the north-west. On the 28th of July, 1848, I saw a flock of them, and killed one. They were very numerous at Chief Justice Butterfield's pond, on the 4th of August, 1848.

SOLITARY SANDPIPER (*T. solitarius*). This bird (so very like our British *T. glareola* (wood sandpiper), both in appearance and habits, frequenting swamps, filled with stunted trees and bushes), is found every year, from the 20th of August, to the 20th of September, a few stragglers remaining till November. I saw one specimen, the 25th of July, 1848, at Burgess' Point; and on the 7th of April, 1850, there were several solitary sandpipers, in Pembroke Marsh. There were generally very shy.*

SPOTTED SANDPIPER (*T. macularius*). I first found this bird, in immature plumage, on the 20th of July, 1847, at Mangrove Bay, Somerset. They are common on all the shores of the islands. A few remain during the winter, and they were not uncommon in April, 1849. The first I got in adult plumage, was on the 16th of August, 1847; another, on the 17th of January, 1848. They do not breed in the Bermudas.

BLACK-NECKED STILT (*Himantopus nigricollis*). A male specimen was shot by Mr. A. Hinson, on the 3rd of June, 1853; it was wading in the pond, near Warwick Church, where it had been noticed for some days. Mr. Hurdis very kindly sent me this specimen, the only one hitherto obtained, and it is now in my collection.

AMERICAN WOODCOCK (*Rusticola Americana*). A single

* Mr. Hurdis states that two other specimens of this sandpiper were shot, on the 10th and 16th of April, 1850.

specimen was shot near Hamilton, in October, 1842; and one was supposed to have been seen at Hungry Bay, a few years afterwards, by Mr. Fozard.

AMERICAN SNIPE (*Scolopax Wilsonii*). These birds are more or less common during the autumnal migration appearing in the month of October. A few remain till the month of January. Several couple were killed in May, 1847, on their northern migration. On the 13th of October, 1849, an immense number of snipe appeared, and remained for a few days, when they again took their departure to the south. I shot two at Brackish Pond, on the 21st of March, 1850, and saw another on the 16th of April, at the same place. This species has sixteen tail feathers. On the 6th of November, 1848, I killed a snipe with eighteen tail feathers, and a few days before that, one with an imperfect tail of seventeen feathers. What were they? Pembroke Marsh was a sure find for snipe, and I think they were the finest and fattest birds I ever killed.

THE ENGLISH SNIPE (*S. gallinago*). I shot one of this species, on the 24th of December, 1847. It precisely answered to the description of the S. gallinago, and had but fourteen tail feathers. I shot another specimen, also with fourteen tail feathers, on the 29th of December, 1847 Both these birds I got in Pembroke Marsh; and there cannot be the least doubt about the bird, and the propriety of adding it to the Bermuda list.

BROWN SNIPE (*Macrorhamphus griseus*). First shot by Captain Orde, on the 29th of September, 1847, at Harris's Bay. Another specimen was killed by Mr. C. Fozard, on the 21st of August, 1848. I never met with this bird myself in the Bermudas.

SEMIPALMATED SANDPIPER (*Tringa semipalmata*). From the 1st of August, to the beginning of November, small flocks of these birds are always to be found at Mangrove Bay, Somerset, and most of the other sandy bays along the coast; also at the Flatts, and Chief Justice Butterfield's pond.

THE STINT OF AMERICA (*T. pusilla*). Arrive about the same time as the former species. On July 31st, 1847, I shot several at Mangrove Bay; and on the 4th of August, 1848, they were in hundreds, at Chief Justice Butterfield's pond, along with the ring plover, semipalmated sandpipers, and tattlers.

SCHINZ'S SANDPIPER (*T. Schinzii*). I first noticed this bird, October 5th, 1847, on Long-bird Island, and killed several at Mangrove Bay, Somerset, afterwards; they were generally in company with the pectoral sandpipers.

PECTORAL SANDPIPER (*T. pectoralis*). More numerous than the last-named species. It had not been noticed previous to September 20th, 1847, on which day I killed two at Mangrove Bay, Somerset. They were common after this at all the swampy ponds and bays, particulary at Mangrove and Shelly Bays, the Flatts, and Chief Justice Butterfield's pond. On the 9th of October, 1849, they appeared suddenly in thousands, particularly at St. George's, after a heavy gale of wind; the parade ground, at that place was swarming with them, and I think Colonel Drummond killed some thirty or forty couple before breakfast; but, with the exception of a few stragglers, they were all gone by the following day.

LONG-LEGGED SANDPIPER (*T. himantopus*). Only one specimen has been observed, which was shot by myself, at

Mangrove Bay, on the 2nd of August, 1848. I killed another, a few days afterwards, at Chief Justice Butterfield's pond, but it unfortunately fell into the middle of the swamp, and I could not possibly get it out.

THE LANDRAIL OF EUROPE (*Crex pratensis*). On the 25th of October, 1847, when out shooting in the dusk of the evening, in Pembroke Marsh, my good old dog "Flora" pointed, and a well-known bird took wing, which I most fortunately killed, and it proved to be a young male landrail of the year. I sent the specimen to the late Mr. Yarrell, and at the sale of his effects, it was purchased by Colonel Drummond, who recognized the skin. The occurrence of this bird in such an "*out-of-the-way place*" as Bermuda, and so far to the westward of its line of migration, is most wonderful; and it certainly gave me more pleasure to find this single bird, than the whole of the other birds put together. I sent a notice of the occurrence of this bird to the Zoologist, in 1849.

CAROLINA CRAKE GALLINULE (*Ortygometra Carolinus*). These birds regularly visit the Bermudas, arriving about the beginning of September. The first specimen I got was at the Sluice Ponds, on the 3rd of September, 1847; it was settling on a branch of a mangrove tree, about four feet from the ground. A few remain throughout the winter. In October, 1849, they arrived in immense numbers. I killed one, January 17th, 1849, at the Sluice Pond; and Mr. Hurdis killed one on the 26th April, 1849, at Warwick Pond. I also found a great many of them, in Pembroke Marsh, my dog pointing them like quail. They were very good to eat, particularly on their first arrival.

YELLOW-BREASTED RAIL (*O. noveboracensis*). Of this

elegant species of crake, two specimens were shot by myself, one on the 5th of October, 1847, and the other, a few days afterwards. Both were killed in the Pembroke Marshes.

LEAST CRAKE GALLINULE (*O. Jamaicensis*). I found this diminutive crake, for the first time, in my favourite shooting quarter, "Pembroke Marsh," on the 19th November, 1847. I saw another at the same place, October 17th, 1848, but lost it, not having my dog out with me. I saw another, a short time afterwards, at the Sluice Ponds; and Mr. Hurdis killed one on the 10th November, 1851, at Pembroke Marsh.

VIRGINIAN RAIL (*Rallus Virginianus*). A male specimen was shot by Mr. Hurdis, in the marshes leading to Spanish Point, on the 6th of November, 1851; and he very kindly sent me the bird, along with the notice of it. This is the only genuine *Rail* met with; a singular circumstance, when we bear in mind that *all* the gallinules, and crake gallinules, known to the continent of North America, have been obtained in the Bermudas.

THE COMMON GALLINULE, or MOOR HEN (*Gallinula galeata*). Not uncommon, and breeds. I obtained several specimens of it, during my stay in Bermuda.

PURPLE GALLINULE (*G. martinica*). Several of these beautiful birds were obtained in April, 1850. One on the 15th of April, which is now in my possession. One caught on the breakwater, at the dockyard, Ireland Island, in April, 1849, is now in the Rev. H. B. Tristram's collection. I saw several, during the same month, at Chief Justice Butterfield's pond, but they were so extremely shy, I could not get near them.

THE COOT (*Fulica Americana*). A few specimens are generally obtained in November and December; and Mr.

Hurdis mentions one as having been shot at Somerset, on the 28th of May, 1847.

Grey Phalarope (*Phalaropus lobatus*). Two specimens only were obtained; one was found dead, floating in Riddle's Bay, March 21st, 1848, and is supposed to have struck the iron lighthouse during the night; it was a male bird, and is now in the Rev. H. B. Tristram's collection; the other, a female, partly in ruddy plumage, was found swimming in Hamilton Water, March 22nd, 1848, and killed by Mr. F. Trimmingham, *with a stick*, and is now in my collection.

Snow Goose (*Anser hyperboreus*). On the 19th of October, 1848, Mr. Hodgson Smith shot two of these birds, in their young plumage, at Riddle's Bay; but, unfortunately for the ends of science, they furnished the dinner-table instead of the cabinet.

The Shoveller Duck (*Anas clypeata*). A single female specimen was shot, in December, 1844, by Mr. C. B. Fozard.

Gadwall (*A. strepera*). Mr. Hurdis kindly favoured me with the following notice on this bird: "A female specimen was purchased from Astwood, the Salt Kettle ferryman, on the 20th of December, 1849; it recovered from its wounds, became quite domestic, and associated with some tame ducks. During the year, 1850, it laid four nests of eggs, small in size, and numbering from twelve to sixteen each time. She was a determined sitter, but her eggs were always bad, and had to be destroyed. In May, 1851, this duck suddenly disappeared, and is supposed to have been stolen from an adjoining marsh. I closely studied the character and habits of this duck, whilst in my possession; and as there cannot be any doubt of its being the female gadwall, I have added it to the list of birds found in

Bermuda." Mr. Hurdis has still some of the eggs of this bird, and I saw several of them myself, in 1850.

DUSKY DUCK (*A. obscura*). Occasionally met with in winter. I got one specimen, on the 19th of December, 1849, from Mr. Fozard, who shot it in Pembroke Marsh, and at the same time bagged *three tame ducks*, with which it was in company. I have seen them several times at the Ponds, near Harrington Sound, but never succeeded in killing any of them.

PINTAIL DUCK (*A. acuta*). The first specimen of this duck was found by myself, in Pembroke Marsh, on the 4th of November, 1847; it was a young male bird. I shot another, a few days afterwards, in the same place, and two or three others were shot by different people during the winter. They were all young male birds, excepting one, a young female.

WOOD DUCK (*A. sponsa*). A female bird of this species was shot by Dr. Cole (20th Regiment), on the 16th of December, 1846.

THE MALLARD, or COMMON WILD DUCK (*A. boschas*). A female specimen was shot by Mr. (now Captain) Martley (56th Regiment), at Warwick Pond, on the 3rd November, 1854. It was found in company with some tame ducks, always taking wing the moment any one approached the water. It was sent to Mr. Hurdis for inspection, and then to Government House, as a present.

BLUE-WINGED TEAL (*A. discors*). Occasionally met with in autumn and winter. I never succeeded in shooting one, although I have once or twice seen them; but Mr. Hurdis gave me a beautiful specimen, which he shot on the 4th of April, 1850, and which is now in my possession.

GREEN-WINGED TEAL (*A. Carolinensis*). An occasional autumn visitant; in some years more common than in others. I obtained one or two specimens of it during the time I was quartered in Bermuda.

AMERICAN WIDGEON (*A. Americana*). Mr. Hurdis sends me the following note:—"During a violent revolving gale, which visited the Bermudas, on Sunday, October 22nd, 1854, numbers of the duck tribe, including this species, took refuge in the creeks and marshes of the Islands. Several American widgeon were shot by different persons, and brought to me for inspection on the following day. On the 11th of Nov., 1854, I killed a female specimen in the Pembroke Marshes. It is singular that this widgeon was never met with during the previous fourteen years in which I had been resident in those islands."

SURF SCOTER (*Fuligula perspicillata*). One of these sea ducks was killed with a stick in the harbour of Hamilton, on the 8th of January, 1849, by Mr. Richard Dill, whose cook, unfortunately, plucked it instanter. The head, however, was found by Mr. Hurdis, and proved to be of the above species.*

SCAUP DUCK (*F. marila*). Occasionally met with. I shot two female specimens, both in their first year's plumage, on the 8th of January, 1849, at Warwick Ponds; and this was the only occasion that I ever met with them in Bermuda.

GOLDEN-EYE DUCK (*F. clangula*). A male specimen was shot, by Captain (now Major) Bull, (56th Regiment), on the 10th of April, 1854, in Pembroke Marshes.

---

* Mr. Hurdis states that another was shot in the Pembroke Marshes, by Mr. Fozard, on the 17th of October, 1854.

Ring-necked Duck (*F. rufitorques*). A specimen of this duck was captured by Mr. Hurdis on the 13th of November, 1850, in the plumage of the year. He endeavoured to keep it alive, for the purpose of watching the change of plumage, but failed in doing so, a cat having bolted with his prize.

Ruddy Duck (*F. rubida*). A single specimen, shot by Dr. Cole (20th Regt.), in a marsh, near Hamilton, November 24th, 1846. It was a young male bird.

Canvass-back Duck (*F. valisneriana*). Mr. Hurdis purchased a specimen of this duck from some boys, by whom it was captured alive in a marsh, near James' Cottage, on the 30th of October, 1851. It was destroyed by ants soon afterwards. On the 23rd of November following, he observed a very fine specimen in White's Marsh.

Buffel-headed Duck (*F. albeola*). One of these ducks was shot in the Pembroke Marshes, by Captain Lyc (20th Regiment), in December, 1845; and though occasionally observed in the Bermudas, subsequent to that period, no other specimens were obtained.

Hooded Merganser (*Mergus cucullatus*). One of the men, on board H.M.S. *Scourge*, Captain Wingrove, caught a female specimen near Ireland Island, on the 10th of January, 1849; the only one hitherto found in the Islands.

Horned Grebe (*Podiceps cornutus*). Dr. Cole shot one of these grebes on the 24th of November, 1846, which is now in the Rev. H. B. Tristram's collection.

Pied-billed Dabchick (*P. Carolinensis*). Two specimens were killed, in October, 1849, and another by myself, on the 5th of February, 1850, in one of the ponds near Harrington Sound.

Brown Pelican (*Pelecanus fuscus*). One of these birds was shot at Hungry Bay, many years ago; and another was killed near St. George's, in April, 1850, which was given to me by Colonel Drummond.

Double-crested Cormorant (*Phalacrocorax dilophus*). The first notice of this bird was by Captain Orde, who shot it at Pitt's Bay, on the 10th of October, 1847; and on the 8th of February, 1848, when sailing past Grace's Island, I shot another. Both these birds were in winter plumage. No other specimens were obtained during my residence in the Islands.

Booby Gannet (*Sula fusca*). Only one specimen has been obtained: this bird flew into one of the soldiers' barrack rooms, at Fort Catherine, on the 3rd of October, 1847. It was beautifully preserved by Colonel Drummond, and is now in my collection.

Frigate Bird (*Tachypetes aquilus*). On the 27th of September, 1848, I observed a large bird soaring about in the dockyard, at Ireland Island, which at last flew into one of the barrack rooms of the Royal Artillery, at the Commissioner's House, but made its escape. After watching it for a long time, I fortunately managed to shoot it, and it proved to be a very fine male specimen. On the 30th of September, 1848, as I was starting for Hamilton, from the dockyard, in my little yacht, "The Petrel," another frigate bird flew over, and close to the truck of my mast; my gun was soon brought into play, and I killed it. The guard at the dockyard was just being relieved at the time, and the bird dropped exactly between the two officers, Lieutenant Fraser, and McNish, of the 42nd, who were standing in front of the men; neither, however, now live to remember the circumstance, as Lieutenant Fraser fell in the Crimea, and McNish was, unfortunately, drowned at Gallipoli. The bird was a female specimen, measuring eight feet six inches from wing to wing. A strong easterly gale had been blowing for some days. One or two others were seen afterwards, but none of them were shot.

TROPIC BIRD (*Phaëton æthereus.*) Very common. These birds arrive regularly every year from the south in March and April. I observed them as early as the 10th March, in the year 1848, and until 25th September of the same year; and again on 1st March, 1850, I saw eight on the north shore, near the light-house. I also observed one on the 19th November, 1849, about twenty miles out at sea. They breed in holes in the rocks on the various islands, and particularly along the south shore, and Gurnet-head Rock, about the beginning of May. The parent birds sit so close that they allow themselves to be caught by the hand; they however show some fight, seizing your fingers in their powerful serrated bill, and occasionally biting very hard. The young birds are marked on the back and wings with transverse bracket-shaped bars of black or brown, but wanting the two elongated centre tail feathers. It is very singular that the young birds are never seen after they leave the holes in which they were reared, and I presume they at once proceed to sea with the parent birds. The Phaëton lays one egg only, of a chocolate colour, with large brown patches, and spotted with black and brown, exactly resembling in colour the egg of our British kestril, only larger, and of a more oval form.

ROSEATE TERN (*Sterna Dougallii*). This was the first bird new to the Bermuda list, that I found in the Islands.

On the 30th of April, 1847, I shot several specimens at Spanish Point. They appear about the end of April, and are very common at Spanish Point, and in Castle Harbour, where they breed. I got their eggs at Gurnet-head Rock, on the 17th of June, 1848. I have also seen a few roseate terns at the North Rock.

COMMON TERN (*S. hirundo*). I have seen a few of these terns at Gurnet-head Rock, but only succeeded in shooting a couple of them near the entrance to St. George's harbour, on the 17th of June, 1848; the only specimens hitherto obtained. They breed along with S. Dougallii, in Castle Harbour.

SOOTY TERN (*S. fuliginosa*). Dr. Cole shot a specimen of this tern in October, 1846. During the whole time I was quartered in Bermuda, I only saw one of these birds, and that in the year 1848: I was walking on the Sand Hills, and saw a bird apparently dead on the ground; I put down my gun and picked the bird up, and was just putting him carefully in paper, when my prize thought fit to come to life and flew away, taking me so much by surprise that I never thought of using my gun. It was a most beautiful specimen, and must have been driven on shore by some heavy gale.

NODDY TERN (*S. stolida*). One example only is recorded; killed by Capt. Tolcher (56th Regt.), near Ireland Island, on the 12th of September, 1854.

SABINE'S GULL (*Larus Sabini*). A single specimen was shot by Colonel Drummond, near St. George's, but the date I do not recollect.

KITTIWAKE GULL (*L. tridactylus*). I found the first specimen of this gull near Spanish Point, on the 5th of

January, 1848, in immature plumage. Colonel Drummond shot two near St. George's on the 4th of February following; one, in adult plumage; the other, in the plumage of the young. I shot another on the 11th of February, 1848, at Spanish Point, and several appeared for a few days in March, 1849, in Hamilton Water.

BONAPARTE'S GULL (*L. Bonapartii*). I noticed this little gull in the Great Sound, on the 23rd of January, 1849, and succeeded in shooting it on the 27th of January during a strong northerly gale. I saw one on the 15th of December, 1849, and another was killed by some one on the 24th of February, 1850, which last I did not get till the 27th of the same month, then too far gone for preservation.

AMERICAN GULL (*L. zonorhynchus*). Rare; one specimen only having been killed by myself near the Dockyard, Ireland Island, on the 1st of January, 1849, during a gale from the north-west.

WESTERN GULL (*L. occidentalis*). A few specimens have been killed; the first, by myself on Darrell's Island, November 17th, 1847; another, 12th January, 1848, both in their brown winter plumage. One or two others were shot near St. George's. I have seen them on several occasions sitting on the rocks at Spanish Point, The Ferry, Pearl Island, and St. David's Head.

HERRING GULL (*L. argentatus*), I first noticed some of these gulls sitting on Long Bird Island, February 23rd, 1848, when I killed four of them, and on the 19th of March following, a few others were obtained.

WANDERING SHEARWATER (*Puffinus cinereus.*) Mr. Hurdis sends me the following note: "Mr. Downes, of the Commissariat, presented me with a living specimen found lying on the high road on the opposite side of Hamilton Water, on the 2nd of June, 1851 ; it was uninjured, and in perfect plumage. On the same day a second specimen precisely similar, was brought to my office by a man who had observed it swimming near the shore opposite Hamilton. By sending his son into the water, and diving below it, the bird was captured by the boy. These are the only specimens I met with in Bermuda."

DUSKY SHEARWATER (*P. obscurus*). This bird breeds on Gurnet-head Rock, and is supposed to be the "Cahow," spoken of in Capt. Smith's account of the Bermudas, 1629. It was found by Captains Orde and McLeod, together with the eggs and young. I have a beautiful specimen, given to me by Capt. McLeod, who caught the bird on its nest in May, 1849.

WILSON'S PETREL (*Thalassidroma Wilsoni*). I have often seen these birds flying about near the North Rock, and once or twice inside the outer reefs in stormy weather, but never succeeded in shooting any of them.

The following birds have been observed in Bermuda ; but, as no specimens were obtained, they are placed on the doubtful list :—

HAWK OWL (*Surnia funerea*). Seen by Col. Drummond at St. George's, quite close to him, on a Sunday afternoon, otherwise it would have been shot.

PASSENGER PIGEON (*Ectopistes migratoria*). Seen by Dr. Cole.

PURPLE SANDPIPER (*Tringa maritima*). Seen by myself at the entrance to St. George's Harbour.

GLOSSY IBIS (*Ibis falcinellus*). Seen by Mr. Hurdis.

RED FLAMINGO (*Phœnicopterus ruber*). Seen by Mr. Hurdis.

Canada Goose (*Anser Canadensis.*)

American Swan (*Cygnus Americanus.*) Shot in White's Marsh; the witnesses still living.

Goosander (*Mergus merganser*).

Black-headed Gull (*Larus atricilla*). Seen by myself at Ireland Island. I was fishing from the shore, near Mr. Triscott's house, and it flew close past me.

FURTHER NOTES AND OBSERVATIONS
ON THE
RESIDENT AND MIGRATORY BIRDS OF THE BERMUDAS.

By J. L. Hurdis, Esq., late Controller of Customs and Navigation Laws in those Islands.

White-headed Eagle (*Haliaëtus leucocephalus*). Seen by myself and several other persons.

Peregrine Falcon (*Falco peregrinus*). The falcon here mentioned (see Major Wedderburn's note, p. 25), was in my possession to the close of November, 1850. A few days previously, it made a desperate attack upon a domesticated female gadwall, and was so roughly beaten off with sticks and stones, as to cause its death. At this time it was gradually changing its plumage, from brown, to a light slaty hue.

A falcon of this species alighted in the dockyard, at

Ireland Island, during a revolving gale, on the 22nd of October, 1854, and was so exhausted by buffeting with the storm at sea, as to allow itself to be captured. I saw this falcon a week afterwards, in the possession of Captain Pye (56th Regiment). It was tethered to a large stone, near that officer's quarters. It was a noble specimen of its kind, and had received no injury whatever.

PIGEON FALCON (*Falco columbarius*). Within a period of four years, I examined five specimens of the pigeon falcon, all of which were killed between the 29th of September, and the 12th of January. I am, therefore inclined to consider it as an autumnal visitor.

THE HARRIER (*Circus cyaneus*). Two specimens only have been examined by me, viz.: one, shot by Mr. Pooley (20th Regiment), in the autumn of 1845, which was found to have earth worms in its stomach; and the other, killed by myself, on the 13th of December, 1851.

LONG-EARED OWL (*Otus vulgaris*). This is an irregular winter visitant, met with in some seasons, between the latter part of October, and the commencement of February. It is remarkable that all the specimens were obtained from Gibb's Hill, the highest part of the Bermudas, the iron lighthouse upon which, was first in operation on the 2nd of May, 1846.

BARRED OWL (*Syrnium nebulosum*). Seen by me, on the 2nd of April, 1851.

GREAT AMERICAN SHRIKE (*Lanius borealis*). It was killed on two previous occasions, viz.: by Dr. Cole (20th Regiment), on the 31st of October, 1846, and by myself, on the 23rd of January, 1847; but, from being in immature plumage, *with the four middle tail feathers wholly black,* they were for some time mistaken for the lesser species, L. ludovicianus, of Audubon.

PIPIRY FLYCATCHER (*Muscicapa dominicensis*). Perhaps the readers of the "Naturalist in Bermuda," will pardon the insertion of the following anecdote, concerning the first specimen of this bird observed in the Islands: In the month of March, 1850, a gallant Lieutenant of Her Majesty's 42nd Highlanders, was in the act of shaving himself for the morning's parade, in the airy costume of shirt and slippers only, when his boat boy, full of enthusiasm, rushed into his master's quarters, and announced "a strange bird," on the topmost branch of a neighbouring cedar; with a spirit worthy of Audubon himself, the gallant officer grasped his "trusty double" in one hand, retaining his razor in the other, and with physiognomy well lathered, stalked forth, in the same light marching order, and, from the middle of the public street, triumphantly brought down the only pipiry flycatcher, met with in the Bermudas to that period.*

HOODED FLY-CATCHER (*Myiodioctes mitratus*). Shot by Mr. C. C. Abbott (20th Regiment), on the 30th of March, 1847. The female companion of this bird was observed at the same time.

WOOD THRUSH (*Turdus mustelinus*). This is one of those rare visitors which appeared with the extraordinary flight of yellow-billed cuckoos in October, 1849. A very fine specimen of this thrush was brought to me at the time by Mr. Trimmingham.

---

* The writer was not present at this exemplification of the pursuit of knowledge under difficulties, and is indebted to the brother officers of his gallant friend for the above statement.

Olive-backed Thrush (*Turdus olivaceus*). Met with in October, 1849, and at no other period. This thrush is not described either by Wilson or Audubon, and appears to have been overlooked by those celebrated ornithologists. Gerard, in his account of the birds of Long Island, describes the species under the above name.

Migratory Thrush (*Turdus migratorius*). On the 13th of March, 1850, I fell in with a small flock of these thrushes in a cedar grove, near Chief Justice Butterfield's pond; and the day following, a friend who resided near the spot, sent me a fine specimen. Five years afterwards, I received a specimen from Mr. Hodgson Smith, of Riddle's Bay; this was on the 29th of March, 1855.

Cat Bird (*Orpheus Carolinensis*). This bird is very destructive to grapes, and feeds with avidity on the small capsicum, or bird pepper, the most pungent of its kind, which it plucks and swallows entire.

Yellow-crowned Wood Warbler (*Sylvicola coronata*). Another specimen was obtained on the 5th of April, 1855, and three others were found (April 30th, 1854), in a collection of skins prepared by Captain Tolcher (56th Regiment), who informed me that he shot them at Somerset from a flock of those birds, which numbered upwards of a hundred. The date of this occurrence I have omitted to make note of.

Pine-creeping Wood Warbler (*S. pinus*). On the 5th of October, 1850, the night being very dark, with a pouring rain, many of these birds were captured on the outer ledges of the lantern of the lighthouse, on Gibb's Hill; one of these I obtained. On the 15th of October, in the same year, Captain Drummond observed from his quarters, at Ireland

Island, a flock of small birds coming in from sea. They approached the Commissioner's house from the north-west, about fifty in number, and formed a long line, rather than a compact body, while on the wing. On reaching Ireland Island, the flock wheeled round two or three times, and then settled upon some cedar trees growing inside the keep, from whence Captain Drummond shot two specimens, one of which was in adult plumage. They proved to be S. pinus, and I had the pleasure of examining these specimens some days afterwards.

Blue-yellow-backed Wood Warbler (*S. Americana*). Another specimen was found in a collection of skins made by Lieut. (now Capt.) Clutterbuck (56th Regt.). It was killed at Ireland Island, but at what period was not stated. The box which contained the collection was sent to me in June, 1853.

Black and White Creeping Warbler (*Mniotilta varia*). A very fine specimen was killed at Ireland Island by Mr. Clutterbuck, on the 27th of October, 1852. In October, 1849, two or three of these creepers were observed on the road between Hamilton and St. George's.

Cedar Waxwing (*Bombycilla Carolinensis.*) I examined two other specimens of this bird shot by Major Wedderburn in Mr. Ewing's Cedar grove; one, on the 6th, and the other on the 10th of April, 1850. Another was killed December 2nd, 1851, by Mr. Darrell, at Peniston's Pond, from a flock of twelve birds, and presented to me. Wilson says, the favourite food of these birds consists of cedar berries; if so, they must find an ample abundance in the groves of Bermuda.

Sky Lark of Europe (*Alauda arvensis*). On the 12th of

June, 1850, I received intelligence from a little negro boy, that a strange bird frequented the north hills, near his mother's cottage. He described the bird as being "less than a pigeon, and of a light green colour about the neck." My young informant also stated that about twelve or one o'clock in the day, this strange bird would rise in the air to a great height, making a "curious noise," when it appeared not to know how to get down again, and would at last tumble down like a stone. This was quite sufficient to rouse my curiosity, and being convinced the boy's description applied to none of the native birds, and that I might meet with the Meadow lark of America on those hills, I immediately proceeded to the spot. Calling at the cottage I learned from the boy's mother the usual haunt of the strange bird, and was on the point of leaving, when, to my utter surprise, the well known delightful cadence of the European sky lark burst in full power and sweetness upon the senses. The woman at once exclaimed, "There he is!" and endeavoured to point out the bird as it floated beneath the clouds above us, but the moving speck continued invisible to my sight. All doubts being removed as to the real character of the strange bird, I followed in the direction in which it was seen by the woman to "go down like a stone," and diligently hunted for an hour without finding it. The weather being warm, I seated myself upon a stone, determined to wait until the interesting songster should think proper to take another flight. In a quarter of an hour he rose at a considerable distance, and after pouring forth a flood of melody (which to one who had not heard the sky lark for sixteen years was an indescribable pleasure), went down, in the true sky lark style, on the top of a neighbouring hill. Of course

I followed, and with no little shame for the act, shot the poor bird as it rose in a fallow, a genuine Alauda arvensis, and a total stranger to the shores of America. Whence it came is an interesting problem for the consideration of ornithologists.*

SNOW BUNTING (*Plectrophancs nivalis*). They also appeared in 1849 and 1853. I have noted twelve specimens shot, and one killed with a stone. In December, 1850, a year fertile beyond precedent in ornithological occurrences, flocks of the snow bunting were observed on the hills of Port Royal, near the lighthouse. Those that I examined were in beautiful plumage, and very fat.

RICE BIRD (*Dolichonyx oryzivora*). The "wandering rice bird" is one of the marvels of the American continent. It breeds in Canada and the United States, north of the 40th degree of latitude, swarms in the rice fields of Pennsylvania early in October, proceeding thence to the islands of Cuba and Jamaica, and early in November departing for the continent of South America. In the month of April they are again seen in Jamaica for a few days on their migration to the north. Darwin mentions this bird, as being found at the Gallapagos Islands in the Pacific; it is therefore known to traverse forty degrees of latitude in its wonderful flight from north to south, and vice versa. What its movements may be from November to March inclusive, remain to be proved by future observers.

The rice bird is not known to visit the Bermudas on its vernal flight, although in September and October, it seldom fails of being found there, generally in small flocks, and on

* The specimen was presented to Major Wedderburn.

one occasion in considerable numbers. They frequent the marshes, where the ripe panicles of seed, from the reeds and sedge, offer an ample abundance of their favourite food.

Two male specimens of this bird, purchased in the market at New York, in May, 1855, and brought to England, sung throughout the summer, and retained the varied plumage of the male bird to November following. One of them then died ; the other continued its song through another summer, and during the whole period it was in my possession, it never assumed the brown and yellow plumage of the female, which I had been accustomed to see in the Bermudas ; I acknowledge myself unable to account for this circumstance.

CARDINAL GROSBEAK (*Pitylus Cardinalis*). This beautiful bird is abundant in the Bermudas, being one of the few permanent residents of those islands. It is found in every garden, and in every grove of cedars, which it enlivens by the brilliancy of its plumage, and the cheerful repetition of its short, unassuming, but pleasing song. It builds twice in the year, the first brood leaving the nest early in April, and the second in June.

A favourite sport with the boys of Bermuda, is the capture of the "red bird" (under which name the cardinal grosbeak is known to the inhabitants), in a very primitive, yet never failing trap, constructed of the stems of the common sage bush (*Lantana salvifolia*). The stems are held together by a stout string attached to the lower corners of the structure, and the centre stick of the roof or cover. By twisting the latter round any degree of tension is acquired. Baited with grains of maize, it is seldom the Cardinal Grosbeak will refuse to enter it. The formation of this simple, yet ingenious trap, is more clearly shown in the accompanying woodcut.

Whether the "red bird" does, or does not migrate to the Bermudas, is a question difficult to decide, it being impossible to distinguish newly arrived strangers from the native birds; and yet, when we consider for a moment that the entire group of islands is of comparatively recent date, resting upon an unknown basis, and that they consist of broken shell, fragments of coral, and small sea shells, which have been washed up from the ocean and moved by the force of the wind, or some other power, into their present forms; are we not justified in concluding that the forms of animal life existing upon a spot which has thus risen from the deep, found their way in the first instance, from the coast of America? If such were the case in a remote period, extending beyond the reach of history, there is every reason to suppose that similar movements occur at the present day.

SUMMER RED BIRD (*Pyranga æstiva*). I was present with Major Wedderburn when he obtained his male specimen, on the 19th of April, 1850, and shall long remember the thrilling cheer of exultation with which he cast his hat into the air, and hailed me across Penniston's Pond, to announce his having killed a "new bird."

SCARLET TANAGER (*P. rubra*). It is somewhat remarkable that the occurrence of this gorgeously plumaged bird should be confined to the spring of 1850 and 1851.

The following is extracted from my note book, May 3rd, 1851: "A splendid male Tanager was recently observed near my own garden; another at St. David's Island by Capt. Drummond, and a third frequented the garden of Mrs. Wainwright, near the town of Hamilton, where it was undisturbed, and from which it suddenly disappeared, after a stay of three or four days. In the early part of the same month, the surgeon of H.M.S. 'Wellesley,' shot a male specimen of this bird near the Admiral's residence."

AMERICAN GOLDFINCH (*Carduelis tristis*). Mr. W. Joel, who resided on the margin of the Devonshire cedar swamp, informed me that several "yellow birds," so termed in the United States, visited that neighbourhood in March, 1850. He described them as small birds, of a bright yellow colour, with black wings. These were doubtless the American Goldfinch (*Carduelis tristis*) of Audubon. I searched for them in vain, and they are consequently unnoticed in Major Wedderburn's list.

LESSER REDPOLE (*Linaria minor*). In March, 1850, flocks of this bird appeared in the neighbourhood of St. George's, where several specimens were obtained. They feed on the ripe berries of the sage bush.

AMERICAN CROW (*Corvus Americanus*). A fine specimen, shot at Spanish Point, was presented to me by Capt. Fayrer,

R.N., in December, 1846, which measured twenty inches in length, and thirty-seven in extent.

In August, 1854, eleven of these crows were observed associating together at Gibb's Hill, by the late Col. Oakley (56th Regt.) This was double the number which had hitherto frequented those parts, and arose, doubtless, from the young of that season.

In an account of the Bermudas, published by the Governor of Virginia, in the year 1623, crows are mentioned, with various other birds, as being found on those islands. "These crows" the writer observes, "direct their flight towards the north-west, at sunset, which makes many conjecture that there are some more islands not far off that way;" and, in another part of his book, he says, "this, with divers other reasons, caused Master More to go out to sea, to see if he could discover any other islands, but he went not far ere ill weather forced him to return."

It is evident, from this statement, that the crow was a native of the Bermudas at that early period, and from the gregarious habits of the bird towards evening, its species (*Corvus Americanus*), is placed beyond a doubt.

I have already remarked in my observations upon the Cardinal Grosbeak, that all the birds inhabiting the Bermudas must, originally, have found their way over sea, from the continent of America, as a heap of broken shell, &c., washed up from the ocean, could have no productions of its own, in the early period of its formation.

Blue Bird (*Sialia Wilsoni*). Although this beautiful and familiar bird appears to be a permanent resident in the Bermudas, vast flights of them, sometimes, arrive from the American coast. This was particularly the case, as observed by Major Wedderburn, in the winter months of 1848. In December, 1849, I fell in with a large flock of these birds in Paget parish. There was an appearance of wildness and vigour about them, which convinced me they were strangers. A small party of eight or ten birds of a different species was observed in the midst of these blue birds, moving with the flock from place to place. I contrived to get within range of the small party, and brought down one specimen, which proved to be a cedar waxwing (*Bombycilla Carolinensis*), in beautiful plumage, but wanting the waxen appendages to the secondaries. It was, consequently, a young bird of that year. Now, the cedar waxwing is a rare visitant in the Bermudas, and is never known to breed there. These cedar birds, then, must have arrived recently in those islands, and in all probability had traversed the ocean in company with the flock of blue birds they were associating with.

There is reason to believe that numbers of the native blue birds leave the Bermudas with these large migratory flocks, thereby causing, as in the summer of 1851, a comparative scarcity of that bird: indeed, but for this supposed movement, it would be difficult to account for the annual increase of the native birds.

YELLOW-BELLIED WOODPECKER (*Picus varius*). There can be no doubt of the migratory habits of this species of woodpecker. In the Bermudas it is a rare and solitary bird, seldom seen except in the winter months. In the early part of April, 1850, as observed by Major Wedderburn, a number of these birds appeared in all parts of those islands. I examined six specimens obtained on that occasion, which were all very fat, notwithstanding the distance they must have travelled. They were all in female plumage, *i.e.*, they wanted the crimson on the throat. This vernal visitation was probably caused by a revolving gale which passed the Bermudas on the night of the 31st of March. H.M.S. "Wellesley," which arrived in the Bermudas on that day from the West Indies, was reported to have met with a great many woodpeckers on the voyage, many of which were captured by the seamen. They were described as having crimson heads and yellow under plumage.

Mr. Gosse found this woodpecker in the island of Jamaica, during the months of December, January, and February only, from which he concluded that it was migratory. It is a summer resident in Prince Edward Island.

YELLOW-BILLED CUCKOO (*Coccyzus Americanus*). The extraordinary flight of cuckoos to which Major Wedderburn alludes, was observed immediately after a strong gale from the south-west, accompanied by torrents of rain, which continued throughout the night. Nine specimens of this cuckoo were sent to me by different individuals on the following day. Thousands, "absolutely thousands," of these birds were observed among the cedar trees on various parts of the south shore, from the Commissioner's House, in Ireland Island, to Somerset, Port Royal, Walsingham, St. David's, and Cooper's Islands, and as far north as St. Catherine's Fort. In the course of two or three days not a straggler remained.

From this it would appear that the yellow-billed cuckoo, which Wilson describes as being shy and solitary, is gregarious at the period of its autumnal migration, and that it crosses the ocean on its southern flight.

BARN SWALLOW (*Hirundo rustica*). I can with safety affirm that from October, 1840, to the 12th of September,

1846, not a swallow of any description came under my observation, though I believe they were sufficiently common in Bermuda in September of the former year.

When the swallow does visit the Bermudas on its southern migration, it makes its appearance with wonderful precision, between the 9th and 16th of August, remaining only for a few days. Why these birds should persist in quitting those islands at a season of the year when the heat is tropical, and the supply of insect food abundant, is a query I will not attempt to answer. Were the periodical want of food, as some writers boldly assert, the principal cause of migration, the swallow would not leave the Bermudas in the manner here stated, or return from the sunny regions of the south, where its natural food abounds at all seasons of the year. Twice only have a few swallows been observed on their vernal flight, in April and May.

RUBY-THROATED HUMMING BIRD (*Trochilus colubris*). My endeavours to ascertain the truth of the tradition alluded to by Major Wedderburn, ended in disappointment. The bird seen by Mr. Darrell was described to me as greenish in colour, with the tail—the only part visible at times—tipped with white. I need not observe that this characteristic appertains to the female.

VIRGINIAN NIGHT HAWK (*Chordeiles Virginianus*). When this bird visits the Islands of Bermuda from the north, it invariably appears between the 20th of September and the 11th of October, and on its vernal flight from the south, arrives with wonderful precision between the 23rd and 30th of April.

Gosse mentions this bird as present in Jamaica in the month of September, and again about the commencement of April. He supposes it to winter in Central America. It is very abundant during the summer in Prince Edward Island.

This wonderful bird, then, is known to pass over twenty-eight degrees of latitude, or sixteen hundred and eighty miles, on its migration to and from the north, traversing vast tracts of ocean in its flight, and disappearing in unknown regions of the south, where it is not unreasonable to suppose that it may continue its course to the temperate latitudes of that hemisphere.

CAROLINA LONG-TAILED DOVE (*Ectopistes Carolinensis*). On the 30th of October, 1854, Mr. Walker presented me with a newly-killed specimen of this dove, which had been captured alive by a coloured person, residing near Spanish Point. It was very plump and fat.

GROUND DOVE (*Columba passerina*). This beautiful dove, the smallest of the pigeon tribe, is common in the Bermudas, where it has thoroughly established itself as a native bird. It is generally seen feeding in flocks, of five or six together, in the highways and enclosures. Its nest is a mere platform of twigs, and is generally placed on the lower branch of a cedar tree. It lays two white eggs, and when disturbed, will attempt to lead the intruder from its nest by fluttering along the ground.

AMERICAN PARTRIDGE, or QUAIL (*Ortyx Virginiana*). Although a migratory bird in some parts of North America, the quail has certainly never found its way to the Bermudas, from 1840 to 1855; Major Wedderburn is, therefore, perfectly justified in considering it to be extinct in those islands. That it was by no means uncommon a few years ago, is sufficiently proved by the testimony of persons, still living, who have enjoyed the sport of quail shooting, when four or five brace of those birds was no unusual bag. There can be no doubt that this beautiful bird was introduced into the Bermudas many years ago, the remembrance of which has passed away. It is not mentioned in the early history of the islands.*

White-eyed Greenlet (*Vireo noveboracensis*). This is one of the native birds of the Bermudas, and is found in abundance all the year round. Its habits very much resemble those of a fly-catcher. In September, it delights to feed on the small white berries of the sweet-scented *tournefortia;* and it is also fond of the small fiery capsicum, known by the name of "bird pepper," the pods of which it plucks and swallows entire.

American Golden Plover (*Charadrius marmoratus*). This plover is a distinct species, and must not be confounded with its European congener. It breeds in high northern latitudes, from which it migrates in countless multitudes to the south, crossing immense tracts of ocean in its flight, and exciting the wonder and admiration of all who love to study the marvellous works of the Creator.

It was in the Islands of Bermuda that the movements of these plover first attracted my attention, and from that distant outpost of the North American coast, I carefully noted down, not only what I observed of them myself during many years, but also much interesting information obtained from persons long resident in the West India Islands, and from masters of vessels navigating that part of the globe. From that source I shall now proceed to make a few observations.

---

* See Note, p. 36.

The golden plover of America visits the Bermudas only at the season of its great southern migration. A few of these birds are met with as early as the 1st of September, or, very rarely, a few days earlier, but, as a general rule, these wonderful migrants pass over the Bermudas in large and numerous flocks, between the 10th and 17th of September. Should the weather be favourable at the time, these flocks pass on at a considerable elevation, in a south, or south-easterly direction, their usual form of flight representing a leading cluster, from which trail two, and sometimes three long lines of single file. In vain does the would-be-sportsman watch these passing flocks by day, or listens to the piping whistle of the multitudes, which are distinctly heard moving in the same direction during the still hours of starlight; not a bird condescends to alight on the sea-girt isles, although a distance of seven or eight hundred miles of ocean must have been traversed on the wing to gain their position; onward they go to the southward, over the vast Atlantic, with a still longer flight before them ere they can reach terra firma! How wonderful must be the power of flight, thus to enable mere land birds to make the ocean their highway from one region of the earth to another, without food, and without a resting-place! More wonderful still, that divine impulse under which these feathered legions move, and by which they are guided across this immensity of open sea at the "appointed time."

Hurricanes sometimes rage with fearful violence in the latitude of our West India possessions at this season. In their course to the north, these hurricanes pass to the westward of the Bermudas, sometimes almost within sight of them, darkening the sky with heavy masses of rapidly moving vapour, and causing a heavy rolling swell to break upon their rocky shores. Should an occurrence of this nature take place while the plover legions are on their flight over sea, it becomes impossible for them to proceed. Driven back and harrassed by the terrific storm of wind and rain,—in which doubtless many perish,—the lone Islands of Bermuda become a refuge to numbers of these birds in their distress, and also to numerous flocks of other over sea migrants. The grassy hills of the south shore and Port Royal, the islands in the Sound, and other open spots, teem on these occasions with flocks of plover; and "gunners" of every shade and degree wage war upon the unhappy fugitives. Many, of course, fall victims to this persecution, (and these are remarkable for being extremely fat,) but no sooner does the weather become settled and fair, than the strangers take their departure for the south, leaving very few stragglers behind them. It is, therefore, as Major Wedderburn correctly states, only in wet and tempestuous weather, that this plover visits the Bermudas in any considerable numbers. A few flocks, probably later arrivals, are met with to the middle of October.

I will now risk the imputation of being tedious, by quoting the following testimony from my notes.

September 11th, 1846. The Schooner "G. O. Bigelow" arrived this day from Halifax, Nova Scotia. The master, Edwin Jones, informs me that while off the east end of these islands, yesterday, hundreds of flocks of plover were seen passing over the vessel to the southward, and numerous flocks could be heard passing in the same direction during the night.

October 18th, 1848. J. R. Place, master of the schooner "Norman," just arrived from Halifax, Nova Scotia, informs me that on his passage hence to that port, about the 12th of September, ultimo, in long. 63·30 W, and lat. 37., he fell in with "vast numbers" of plover, in flocks numbering from thirty to a thousand each, all flying due south by compass; weather moderate at the time. He furthur states that he could hear numerous flocks passing over his vessel in the night of the above-mentioned day, that he does not think there was anything easterly in the flight of these birds, and that he expected to hear that Bermuda had been teeming with them. This flight, however, did not pass over or visit the Bermudas.

September 5th, 1849. Mr. Samuel Nelmes, one of the oldest sea-going navigators of these Islands, told me to-day, as he cleared out his schooner for Prince Edward Island, that, when commanding the brigantine "Carib" twelve or thirteen years ago, on a voyage from London to the Bermudas, and sailing in the latitude of those Islands, in the month of September, with fair breezes and a continuation of remarkably fine weather, the vessel sailing four and five knots only, and the Bermudas distant between four and five hundred miles, he fell in with endless flocks of plover, all flying in a south-east direction.

On referring to the Custom House records, I found that this voyage was performed in the year 1833, the date of the master's report at the Custom House, in Hamilton, being the 14th September. It is therefore evident that the vast flight of plover he alludes to must have crossed the latitude of the Bermudas on or about the 10th of September, at the distance of upwards of a thousand miles from the nearest part of the American coast.

November 25th, 1851. Edwin Jones, master and part owner of the schooner "G. O. Bigelow," when reporting his vessel from England this morning, informed me that about nine or ten days after his departure from Bermuda in September last, with pardoned convicts on board, for England, the weather for several days having been light, with an easterly breeze, and the vessel being then between five and six hundred miles east of these Islands, great multitudes of birds, which he took to be plovers, were observed passing over the vessel in a southerly direction, for two days. These birds he describes as flying in flocks, some of which amounted to many thousands, others to considerably less, diminishing in number to parties of fifty and thirty. He also states that during the whole of the intervening night, these flocks could be distinctly heard passing over the ship.

The "G. O. Bigelow" left the Bermudas on the above mentioned voyage, on the 3rd of September, 1851, consequently, this immense flight, doubtless the golden plover of America, must have been observed by him about the 12th of 13th of that month.

These great migratory columns of plover do not appear to cross the ocean, to the westward of the Bermudas, a few flocks only being met with in that direction by the numerous vessels which trade between those Islands and the United States.

Let us now endeavour to trace the southern destination of these wonderful birds.

A highly respected friend, long resident in the Island of Antigua, assured me that, in the month of September, that island was annually visited by countless flocks of plover, which arrived from the north, and were precisely similar to the golden plover which visits the Bermudas. My informant stated, that on one occasion, when the weather was dark and stormy, these plover made their appearance in such multitudes at St. John's, the chief town of the colony, that the inhabitants were seen, in every direction, shooting them from their doors and windows; indeed, so numerous were they, that boys destroyed them with sticks and stones, and shooting them soon ceased to be considered sport. He added, that in ordinary seasons, the plover are not seen in such immense numbers, although they never failed to be very abundant. They remained in the island for ten or fifteen days only, taking their departure as soon as the weather became settled.

Another gentleman, who had resided in Martinique, gave the same account of the golden plover in that island, stating, at the same time, that it was impossible to exaggerate the numbers which sometimes appeared there.

Sir Robert Schomberg mentions, in his History of Barbadoes, that during a south-west gale which prevailed at that island on the morning of the 12th of Sept., 1846, the flights of wild birds were so numerous, that they were struck down with stones, and thousands were shot, and the Barbadoes newspapers asserted that there had not been so great a flight since the storm of 1780. What these "wild birds" were, the author does not state, but there can be no doubt that he alludes to the golden plover of America. Now, I have already shown that an immense flight of these birds passed to the eastward of the Bermudas on the 10th of that month, and during the ensuing night. If, then, we suppose these birds to travel at the moderate speed of thirty or thirty-five miles per hour only, over the eleven hundred and fifty miles of ocean which intervenes between those islands, we shall have no difficulty in accounting for their appearance at Barbadoes on the morning of the 12th of September. Of course, I do not pretend to assert the identity of the two flights, although it must be admitted that the evidence leans strongly to that conclusion.

Whence these flights of plover proceed, after visiting the West Indian Archipelago, is unknown. There can be little room for doubt, however, that the shores of Venezuela and Guiana, and the interior of the South American continent, will prove to be their destination.

It is singular that the course by which this plover returns to its northern haunts, should also be unknown at the present day. Some writers affirm that it passes rapidly through the northern parts of the United States, in small flocks, in the spring of the year; this, no doubt, is correct with reference to a few, though it still leaves the vernal migrations of the millions involved in mystery.

Sir William Jardine mentions this bird as being found in Australia and Hindostan. Let us hope that some observant naturalist will favour us with the particulars of its history in those possessions, whence it comes, and whither it goes, for assuredly it must be migratory in those countries, as elsewhere.

KILDEER PLOVER (*Charadrius vociferus*). This is the latest of the plover tribe which visits the Bermudas, appearing from the north in the months of December and January. It is found in small flocks, and is conspicuous, from the beautiful black band across the breast. Its note is peculiarly soft and pleasing. It is not seen in the spring.

American Ring Plover (*C. semipalmatus*). I have met with this species from the 16th of August, to the latter part of October, when it disappears on its southern migration.

Piping Plover (*C. melodus*). On the 22nd of September, 1850, two of these very pretty dove-coloured plover were observed feeding on the rocks in Hamilton Harbour. They had a very soft and musical note.

Black-bellied Plover (*C. helveticus*). The specimen killed by Major Wedderburn was an adult in full plumage.

That the young of this species is frequently mistaken for the golden plover there can be little doubt. In looking over what were supposed to be fine specimens of the latter, their superior size, viz.: eleven and sometimes twelve inches in length, the larger head, and the mottled black of the under plumage, have inclined me to believe that they belonged to the black-bellied species. The length of *C. marmoratus*, may be set down at ten and a half inches for the larger specimens, and ten inches for the smaller ones.

Sanderling (*Tringa arenaria*, of Audubon). In October, 1854, I found two specimens of this bird in a collection of skins preserved by Capt. Tolcher (56th Regt.), both of which had been shot in the Bermudas.

The sanderling has been found in those islands from the 4th of September to the 10th of November.

Great Blue Heron (*Ardea Herodias*). This wary bird is not unfrequently met with on the shores and creeks of the Bermudas. I never observed it for several years, without noting the circumstance, and have no doubt in my own mind, of its migratory habits. It is first seen about the 19th of September, continues to arrive in the month of October, when it is sometimes rather common, and is only occasionally met with from that period to the end of April, it then disappears for the summer months; thus closely resembling the belted kingfisher in its movements.

A number of these herons sought refuge in the Bermudas during a violent gale, which swept over those Islands on the 14th and 15th October, 1848; they appeared on all parts of the coast, from St. George's to Ireland Island, and, from the circumstance of five being captured alive in different localities, it would appear that they landed in a state of great exhaustion. A friend, stationed at Ireland Island, witnessed the arrival of two herons of this species during that storm; they made the land from the north, flew close to the surface of the waves, and appeared to be much fatigued.

BLUE HERON (*A. cœrulea*). Of seven specimens of this heron, which came under my observation, four were shot in April and May, and three in September and October. It may, therefore, be considered both a vernal and autumnal visitor to the Bermudas. Three of the spring specimens were beautiful exemplifications of the change from the white plumage of the young, to the rich vinous purple of the adult bird.

AMERICAN BITTERN (*A. lentiginosa*). In the stomach of one, shot in the Pembroke Marshes, was found an eel, six inches long, a mouse, a dragon fly, a grasshopper, and part of a small golden carp.

LEAST BITTERN (*A. exilis.*) Another specimen of this miniature bittern was captured alive, by a whale-boatman, on the rocks of St. David's Head, on the 20th of April, 1853, where it must have alighted on its flight to the north. There can be no doubt of the migratory habits of this bird.

BLACK-CROWNED NIGHT HERON (*A. nycticorax*). Twelve specimens, which I examined in the Bermudas, were all in the spotted plumage of the young; one of these, shot on the 9th of February, had the irides bright carmine, and the long plumes at the back of the head beginning to appear.

These birds vary much in size.

HUDSONIAN CURLEW (*Numenius Hudsonicus*). Of four specimens which came under my observation, one was shot on the 14th of August, and the remainder in September.

GREAT YELLOW-SHANKS TATLER (*Totanus vociferus*). The only instance of the vernal appearance of this bird, occurred on the 5th of June, 1852, when a single specimen was killed at Hungry Bay. Like all other birds which visit the Bermudas at that season, it was in beautiful plumage.

SPOTTED SANDPIPER (*Totanus macularius*). In 1849, a few of these sandpipers visited the Bermudas between the 14th and 28th of April, being the only instance on record of their being seen at that season. In the autumn it is the earliest of the migrants from the north; and I have reason to believe that a large flight of them passed over the Bermudas on the 31st of July, 1850: The evening was clear and tranquil; their note of *tweet, tweet,* was distinctly heard for full fifteen minutes, and yet the birds were invisible, even with the aid of a telescope.

This little sandpiper swims and dives with great ease; one which I wounded, fell into the sea, and, on being approached, dived nearly to the bottom of the clear water, which was upwards of three feet in depth; this it repeated two or three times; and, on rising from one of these descents, near the rock on which it stood, it was captured.

BLACK-NECKED STILT (*Himantopus nigricollis*). My

note-book has the following: "June 3rd, 1853,—Mr. A. Hinson sent me a very fine specimen of this bird, which he had shot about an hour previously, in the pond near Warwick Church. This bird had been noticed wading about that piece of water for two or three days, and was killed at my particular request. It is the first instance on record of its having been met with in the Bermudas. It measured $13\frac{5}{10}$ inches in length, by $26\frac{6}{10}$ inches in extent; bill, to the gape $2\frac{7}{10}$ inches; naked portion of the tibia, 3 inches; tarsi, $4\frac{4}{10}$ inches. In the markings of the plumage, &c., this curious bird agreed perfectly with Wilson's account, excepting that the darker parts appeared to be of a *black-purple*, and the tail, *a light-drab*, glossed with the same black-purple towards the extremity.

It proved, on dissection, to be a male bird, and was extremely fat."

American Snipe (*Scolopax Wilsonii*). Our limited knowledge of the geographical distribution of this bird compels us to suppose that it is confined to the continents of North and South America. Breeding in high northern latitudes, it migrates in immense numbers to the south; and that many of these traverse the ocean in their flight, is proved by the regularity with which they visit the marshes of Bermuda.

In the month of September, when the heat is still tropical, the snipe sometimes appears in those islands as early as the 13th of that month; in other seasons they are not met with until the commencement of October. A few couple are sometimes killed in April and May, but during the summer months it is never seen. In the season of 1846—7, a hundred couple of snipe were shot by the officers of the 20th Regiment, stationed at Hamilton, and, in that of

1849, my own chasse amounted to fifteen and a half couple.

Many of these birds appear to winter in Jamaica, being found in that island from October to April. Gosse alludes to an instance of twenty-two couple being shot there in one day; and Dr. Von Tschudi, in speaking of the zoology of the neighbourhood of Valparaiso says, "the snipes found in the little plain between the bay and the lighthouse, are in colour precisely like those of Europe, from which, however, they differ by having two more feathers in their tails." A more perfect description of Scolopax Wilsonii could not be desired.

CAROLINA CRAKE GALLINULE (*Ortygometra Carolinus*). This bird is one of the marvels of American ornithology. Wilson states that its history is involved in profound mystery, inasmuch that no one knows from whence it comes, or where it goes. He then quotes two instances of the bird being met with at sea, at distances of one hundred and three hundred miles from the American coast, from which he concludes that the great body of those birds must winter in countries beyond the United States, and that Heaven has gifted them, "in common with many others," with instinctive judgment and strength of flight sufficient to seek a more genial abode during the winter season.

Taking this luminous statement for our guide, let us proceed to inquire into the movements of the Carolina Crake, after its departure from the shores of North America.

Heavy and sluggish as this bird may appear when disturbed in its marshy retreat, there can no doubt that it possesses great powers of wing, and the mere fact of its never failing to visit the Bermudas on its great southern migration, is alone a sufficient proof of its prowess in that respect.

In the month of August I have noted a single instance of its being met with, viz., on the 24th, when a specimen was shot in the Pembroke marshes. In September they are not uncommon, and in October more abundant than at any other period, disappearing in some seasons at the end of that month, while in others, a few remain to the 25th of November, and sometimes even beyond that period.

In the years 1849, 1850-1-2, the Carolina Crake was observed to visit the Islands of Bermuda, on its vernal migration, appearing in the latter part of February and the ensuing months of March and April. Of these, ten specimens were shot, and three taken alive. Of the latter, one flew into an officer's quarters at St. George's, where it was captured; another was found in the barrack pig-stye at the same place, and the third was taken from a cat in my own house.

During a south-west gale which prevailed in the Bermudas on the 9th of October, 1849, thousands of these crakes suddenly appeared in the marshes; and the snipe, (*Scolopax Wilsonii,*) also became unusually abundant about that time. On the 29th of the month not a single crake was to be found; the numerous flight had proceeded on its mysterious journey; we know not for certain in what direction, though we humbly presume to the south.

It cannot be said that this departure was occasioned by the want of food, for the marshes were in splendid condition for birds of this genus; and certainly a temperature, ranging between seventy and eighty degrees of Fahrenheit, could not have caused it.

The Carolina Crake is found in the Island of Barbadoes, and Gosse mentions it as frequenting the swamps of Jamaica; but of the great body of these birds mentioned by Wilson as annually departing from the shores of the northern States, there is little cause to doubt that the rivers and marshes of South America will prove to be their southern haunts.

There is one circumstance connected with the history of this, and, indeed, of many other migrants, which I have not adverted to, viz., the very fat condition in which they arrive in the Bermudas, and which renders it sometimes difficult to skin specimens. Now this extreme condition has always appeared to me to be a provision of nature, to sustain these birds on their long and arduous flight from one region to another; if not, how are we to account for the maintenance of the legions of plover, sandpipers, and other birds which traverse the Atlantic, probably for thirty or thirty five degrees of latitude, without food?

COMMON GALLINULE, or MOOR HEN (*Gallinula chloropus*). This is one of the native birds of the Bermudas, rearing its young in pools and swamps, where the dense growth of flags and sedge renders it almost impossible to follow it.

In October this gallinule is more common, appearing suddenly in marshes and ponds, where for months previously it had been unknown. This autumnal appearance must arise, either from the scattering of native broods, or from an influx of migrant strangers from the American coast. I am inclined to think the latter the most likely cause. In October, 1850, a common gallinule was shot in the waters of Hamilton Harbour, and on the 16th of September, 1854, a boy brought me a living specimen which he had captured in the back yard of his mother's house in the town of Hamilton. Originally this bird must have found its way from the American coast, and we have no reason to suppose that its habits are changed in this respect.

HYPERBOREAN LOBEFOOT, OR GREY PHALAROPE (*Lobipes hyperboreus*). On the 8th of March, 1852, a third specimen was obtained by myself, when taking a walk round the north side of Brackish Pond cedar swamp. On disturbing the bird, its sluggish mode of flight induced me to think it was wounded. I gave chase immediately, and, after two or three flights, succeeded in knocking it down with my walking-stick. It was a beautiful male specimen, and when skinned showed no trace of injury.

It is remarkable that all the specimens were obtained without the aid of a gun.

SNOW GOOSE (*Anser hyperboreus*). A wing, pertaining to one of the specimens mentioned by Major Wedderburn, was fortunately saved by Mr. Smith, which removed all doubts as to the species it belonged to.

In October, 1849, two "white geese" were observed in Mangrove Bay, and on the 10th of March, 1851, four dark grey coloured geese were seen on the wing, near Peniston's Pond. These were doubtless *Anser hyperboreus.*

DUSKY DUCK (*Anas obscura*). Of this duck, I have known nine or ten specimens to be killed in the Bermudas. It appears about the middle of September, and is last seen in December. It is highly esteemed for the table, and from its disposition to associate with the common tame duck, I think it might be domesticated with little trouble.

WOOD, OR SUMMER DUCK (*A. sponsa*). The specimen of this duck mentioned by Major Wedderburn, was killed in the marshes near Hamilton, and being a female in imperfect plumage, I experienced some difficulty in identifying it. Audubon and Wilson are seldom sufficiently explicit in the descriptions of female and immature plumage, but with the aid of De Kay's New York Fauna, I was enabled to ascertain the species. This specimen was remarkable for having the eyelids of a bright yellow colour, and the eyes surrounded by a large white patch; the former characteristic is not mentioned by any author that I am acquainted with.

BLUE-WINGED TEAL (*A. discors*). This beautiful species of the duck genus is decidedly migratory, and not unfrequently visits the Bermudas on its southern flight. About the 20th of September it is first met with, and continues to be seen at different periods until the 24th of December. In the month of October, however, they are most numerous, particularly when a storm is raging, or has passed between those islands and the American coast. On its vernal migration to the north it is very rarely seen, and then only at the end of March, or beginning of April.

Nine couple of these teal were shot in the Pembroke marshes, after the gale of the 22nd of October, 1854, and many more at St. David's Island, where a native sportsman is said to have killed sixteen couple of them during the gale.

SURF SCOTER (*Fuligula perspicillata*). Another specimen of this bird was shot in the Pembroke Marshes, by Mr. Fozard, on the 7th of October, 1854.

SCAUP DUCK (*F. marila*). A scaup precisely similar to those shot by Major Wedderburn, was killed by Mr. C. C. Abbott (20th Regt.), on the 19th of December, 1846. All these birds measured from sixteen to sixteen and a quarter inches in length. The delicate pencilling of the second plumage was beginning to spread in the rich brown of the upper parts, and the white mirror was *not confined to the secondaries*, but extended to the primaries, the four or five external ones excepted. A broad white band of brilliant white plumage surrounded the base of the bill. They were evidently young birds, and probably all females.

Two species of scaup are described in De Kay's Fauna of New York; the larger, *F. marila*, 19 to 19 5 inches in length; and the lesser, *F. minor* (*F. marila*, of Audubon), 17 inches in length, the mirror of which *is restricted to the secondaries only*.

Is the female of the larger species, which I consider the Bermuda specimens to belong to, so much less than the male?

HOODED MERGANSER (*Mergus cucullatus*). Col. Drummond (late 42nd Highlanders) when stationed at St. George's, shot a merganser on the 23rd of December, 1850, which measured nineteen inches in length, and was marked with *white* on each wing. I did not see this specimen, though, from the description, there can be no doubt it was the young of the Hooded Merganser.

HORNED GREBE (*Podiceps cornutus*). On the 1st of February, 1855, Captain Tolcher (56th Regt.), presented me with a beautiful specimen of this grebe, in spring plumage, which he shot near Spanish Point. It was in company with three or four others at the time.

PIED-BILLED DOBCHICK (*Podiceps Carolinensis*). On the 18th of November, 1851, I found the perfect skeleton of a grebe of this species, on the margin of Chief Justice Butterfield's pond.

DOUBLE-CRESTED CORMORANT (*Phalacrocorax dilophus*). In January, 1847, a single cormorant of this species appeared in the waters of Bermuda, and became the object of many a sporting cruise. It continued, however, to baffle all attempts to shoot it, and disappeared early in March.

FRIGATE BIRD (*Tachypetes aquilus*). One of these birds was shot by Captain Clutterbuck (56th Regt.), on the 30th of September, 1852, and another by Captain Tolcher, of the same regiment, on the 2nd of April, 1854.

TROPIC BIRD (*Phaëton Æthereus*). On the rocks and craigs of the distant and secluded Islands of Bermuda, where, doubtless, in ages past, myriads of sea fowl of various descriptions, nestled and reared their young, the tropic bird, two species of tern, and the fabulous "Cahow" of the early settlers, are the only sea birds which now frequent those shores in the breeding season.

The Phaëton Æthereus, commonly known by the name of "boatswain," "longtail," or "tropic bird," is a summer resident in those islands, making its appearance from the south, between the 4th and 18th of March, and taking its departure early in the month of October.

The favourite resort of these interesting birds is among the small islands, at the entrance of Castle Harbour, on the shores of Harrington Sound, and along the south coast from the light-house to the north-west extremity of Somerset. Here, at the early period above-mentioned, and conspicuous by the glittering whiteness of their plumage, and by the two long slender feathers of the tail, numbers of these phaetons may be seen busy on the wing, wheeling occasionally in their flight, and dashing perpendicularly into the blue waves to secure their prey, precisely in the manner of the terns.

On the 10th of May, I have ranged the rugged coast, frequented by these birds, for miles, and found it, to all appearance, deserted not only by them, but by every other species of sea bird, although, on a careful examination, the rocky cliffs were found to abound with the Phaëton æthereus in the act of incubation. Those not immediately employed in this duty, were unquestionably seeking food at a distance in the ocean, but at what period of the day they return to their breeding haunts, I have not been able to ascertain.

The tropic bird makes no nest, but having selected a hole, or cavity in the rock, sometimes elevated, and at others merely beyond the reach of the waves, invariably lays a single egg. Some of these holes are superficial, others appear like rabbit burrows in the softer rock, and, in a few instances, I have found the entrance barely large enough to admit the arm, and too deep to allow of reaching the egg with the hand; indeed, on one occasion, I could only ascertain the presence of the old bird, by touching it with the end of a ramrod, and thus exciting its well-known grating cry.

When a breeding-place is intruded upon, the sitting bird makes no effort to escape, but allows itself to be taken by the hand; not, however, without some resistance from its strong and sharp-pointed bill: both male and female are captured in this manner.

The egg varies considerably in colour, some specimens being of a reddish grey, thickly covered with streaks and blotches of Indian red, deepest at the larger end; others are of a drab colour, finely speckled with the same deep red. The young remain in the nest, or breeding-place, until capable of flight: they are at first covered with long white down, which gradually disappears as the bird advances in growth and acquires its first plumage of white, marked on the back and wings with transverse bracket-shaped bars of black, or dark brown: the tail is rounded, and the elongated feathers of the centre do not appear for some time. The following woodcut represents a young phaeton in the unfledged state.

To give an idea of the number of these tropic birds, which breed along the south shore of Gibbs' Hill, and Somerset, I have only to remark that in 1847, being in want of specimens for some friends, I sent word to the sergeant's party, stationed at the Gibbs' Hill telegraph, (which is close to the light-house,) that I would give a shilling each for a few of those birds in full and perfect plumage. I soon received a message in reply, requesting me to go up and select from a number which had been captured. On my arrival at the station, I found between forty and fifty of these birds confined in a small room, the whole of which, with a number of eggs, had been taken in one day. This was on the 15th of May, at which time the eggs were within a few days of hatching.

From the diminutive size and backward position of the feet, this bird is unable to walk in the ordinary mode, but resting its breast upon the ground, and partially spreading its wings, it contrives to shuffle from place to place in a peculiar and awkward manner. Audubon has described the tarsi and hind toe of this bird as being *yellow;* but in the many specimens which I have examined, these parts were of a dead white, tinted with light blue.

The general length of this bird, including the tail feathers, is from twenty-eight, to thirty-three inches. Many specimens are found with one elongated feather only in the tail; these, of course, are imperfect. The male bird is deeply tinged with a beautiful carmine, or roseate hue.

ROSEATE TERN (*Sterna Dougallii*). The eggs and young of this bird have been obtained at Gurnet Head, or Black Rock, on the 1st of August, from which I infer that it rears two broods in the season. It is not met with during the winter months.

COMMON TERN (*S. hirundo*). This species breeds on the Gurnet-Head Rock, forming a social community with the roseate tern. A friend who visited that rock in the middle of August, assured me that he found it teeming with terns and their young.

These birds are invariably called "redshanks," in the Bermudas, from the colour of their legs. The young, however, have the bill and feet black, or dark brown, the red colour not appearing for some time.

SOOTY TERN (*S. fuliginosa*). A very beautiful full plumaged specimen of this tern was presented to me on the 23rd of October, 1854, by Mr. John Darrell (son of the present Chief Justice), by whom it was found, in the parish of Devonshire, laying on the ground in a state of exhaustion. There had been a severe gale the day preceding.

KITTIWAKE GULL (*Larus tridactylus*). This is one of the few gulls which visit the Bermudas during the violent westerly gales of the winter.

BLACK-HEADED GULL (*Larus atricilla*). A gull of this species, captured alive by a fisherman, in the winter of 1851—2, was confined for some time in a spare room, but eventually effected its escape.

WILSON'S PETREL, or MOTHER CARY'S CHICKEN (*Thalassidromo Wilsonii*). I examined a very fine specimen, shot by Mr. Harford (56th Regiment), on the 30th of June, 1853. It was killed some miles distant from the shore.

GREAT BLACK-BACKED GULL (*Larus marinus*). In looking over my note-book, I find the following gull has been omitted in the general list of birds. On the 24th of December, 1851, I examined a living gull, which had been captured in the Great Sound, a few weeks previously. It was of a darker mottled-brown than any former specimens of the gull family, and measured twenty-four inches in length; tarsus, 2·4 inches; eyes, dark brown; outer half of the bill, black, the inner portion, light horn colour; legs and feet, clay colour; nails, black.

I presume this to be the young of *L. marinus*.

LITTLE AUK, or COMMON SEA DOVE (*Mergulus alle*). One of these birds was captured alive on the 28th of January, 1850, by a servant of the Rev. J. U. Campbell, at Ireland Island. It was in company with four or five others, on a piece of grass land, near that gentleman's house. Unfortunately, this specimen was destroyed by a pig before I had an opportunity of seeing it. My information was obtained from Mr. Campbell himself, who had this bird in his possession.

The Cahow.

Governor John Smith, in his general History of Virginia, published in 1629, containing an account of the Bermudas, then appertaining to the government of that Colony, mentions a mysterious bird, as found in those islands, and there called the "cahow," which was noticed "for the tune of his voice." As the writer's description of this bird is somewhat romantic, I will quote his own words.

"The cahow is a bird of the night, for all the day she lies hid in holes in the rocks, where they and their young are also taken with as much ease as may be, but in the night, if you but whoop and hallow, they will light upon you, that with your hands you may chuse the fat and leave the leane: those they have only in winter. Their eggs are white."

Purchas, who described the Bermudas a century later, viz., in 1738, makes the following statement.

"Birds are equally abundant and various, many of the species peculiar to the Island; the most singular was one called Cowhow, or Cowhie, about the size of a plover, which come forth only in the darkest nights of November and December, hovering over the shore, making a strange hollow, and harsh howling. The most approved mode of taking them was by standing on rocks by the sea-side, whooping, hallooing, and making the strangest outcries, which attracted the birds, until they settled on the very person of the hunter."

Here we have the testimony of two writers distinctly pointing to the existence of a "singular" bird in the Bermudas. Let us examine these statements, and separating truth from fable, endeavour to ascertain the true character of this marvel of former days.

Governor Smith says it is nocturnal in its habits, and noted for the tune of its voice, hiding in holes in the rocks by day, where it rears its young, and is easy to capture; he also describes its egg as being "white," and Purchas further informs us that the bird is about the size of a plover.

Being unable to obtain information from any one regarding the existence of this bird at the present day, I determined to visit the islands in the vicinity of Castle Habour, for the express purpose of ascertaining that point. A friend, imbued with a taste for natural knowledge, joined me on this occasion. Embarking at the Ferry House in a four-oared cutter, on the 28th of June, 1847, we visited several of those islands, and were pleased to learn from persons there resident, that the Cahow was still known by its old name, which was described to us as an imitation of its peculiar note *cao-hoo*, and that it still continued to breed in that locality. A boy assured us that he had recently caught two of those birds on Cooper's Island, and described them as "brown or whitish," and about the size of a duck.

I was particularly anxious to examine the Black, or Gurnet Head Rock, which rises boldly from the sea near the entrance to Castle Harbour; it being a favourite breeding place for the tern family, and where I thought it very probable the cahow might also be found, but a strong breeze caused so much swell outside Castle Harbour, as to render the attempt unsafe.

On or about the 1st of August, in the same year, Mr. Salton Smith, of St. George's, kindly undertook, at my suggestion, to visit the Black Rock, with the view of obtaining specimens of the Cahow. The following account of his proceedings I extract from my notes.

"He tells me, that on the occasion alluded to, the sea was rougher than he anticipated, and that he had some difficulty in landing a boy upon the rock. This, however, he succeeded in doing with the dingy, and the boy twice returned with specimens of young sea birds, consisting of about a dozen "redshanks," (terns,) and two cahows, all which were safely got on board the dingy. He then went for eggs, and returned with some dozens in the fold of his shirt; with these he attempted to jump into the frail bark as it rose upon a sea, but missing his footing, fell into the water, and was in danger of being drowned. Mr. Smith, in his endeavours to save the boy, was carried on the rock, the dingy was upset, and stove, and the whole of the specimens and eggs lost. How the parties got back to their sail boat I did not learn, but they did so, in safety.

Mr. Smith describes these cahows as being two beautiful birds, one of them in particular, which he took to be a male. They were about the size of a common duck, white below, and mottled with dark colour about the head and upper parts of the body. The two were found together in the same hole; the old birds were not seen. This is the full extent of the information gained upon this occasion.

On the 17th of May, 1849, Captains Orde, and McLeod (42nd Highlanders), visited the Black Rock. They landed without difficulty, and upon a ledge about half way from the summit, captured two fine specimens of *Puffinus obscurus*, the dusky shearwater, of Audubon, one of which was sitting upon a single white egg; the other had nothing under it. Both these birds were found in holes of the rock, and allowed themselves to be captured by the hand. A young bird of the same species, covered with black down, was also found upon the rock: it was not disturbed. The egg was nearly as large as a common fowl's, but more finely polished on the surface.

- I carefully examined these specimens, and found the length to be 13½, and 13¾ inches; extent, 26 inches. They proved, on dissection, to be a male and female. This discovery is highly interesting. The larger species, *Puffinus cinereus*, or Wandering Shearwater, of Audubon, is rarely met with in the Bermudas, and certainly does not breed upon their shores. It is, therefore, beyond all doubt that the cahow, described by Governor John Smith, in 1629; by Purchas, in 1738; and by the native islanders of the present day, is identical with *Puffinus obscurus*, or the Dusky Shearwater, discovered on the Black Rock, by Captains Orde and McLeod.

Whether the dusky shearwater continues to frequent the rocky coast of the Bermudas during the winter months, I am unable to state. It is not described by Wilson.

TENGMALM'S NIGHT OWL (*Ulula Tengmalmi*). November 22nd, 1847: Examined an owl, which flew on board the schooner, "Robert," Williams, Master, on the 14th instant, while on her voyage from Halifax, Nova Scotia, to these islands. The vessel was in latitude 41-10 N, and longitude 63-40 W, at the time, and the wind blowing very fresh from the south-west. The owl was alive when landed, but so weak, as barely to be able to stand, and it died the following day.

Length, 8½ inches; wing, from carpal joint, 6½ inches; third and fourth quills, longest. Compared it with Audubon and Yarrell, and have no hesitation in pronouncing it to be a genuine specimen of Tengmalm's Owl. Though not actually a visitant to these islands, it was, doubtless, on its southern flight, when fallen in with by the "Robert," and, but for that occurrence, *might* have been entitled to a place in our list of Bermudian Birds.

BLACK-BILLED WHISTLING DUCK (*Dendrocygna arborea*). This bird is sometimes introduced into the Bermudas from Turk's Islands, where it is found in a state of nature, and reared from the nest as a semi-domestic bird.

Although a native of the Bahamas, no *wild specimen* of this duck has been met with in the Bermudas.

REPTILIA.

GREEN TURTLE (*Chelonia mydas*). This is the common turtle of the Bermudas. Not abundant.

HAWKS-BILL TURTLE (*Chelonia imbricata*). Sometimes called "tortoishell turtle." Not unfrequently taken.

A species of Skink (*Scincus*), is very common on the Islands, frequenting the old walls and stone heaps in the cedar groves. The collector must be careful, in capturing his specimen, not to take hold of the tail, as in the struggle to free itself, this agile creature is apt to save its body at the expense of its appendage.

Mr. Hurdis has the following note upon this species:—"May 8, 1849. Examined a specimen of the lizard common in these Islands, which was found lying dead on a public road. Length 7 inches; upper parts entirely dark brown, *without any appearance of longitudinal stripes;* cheeks, and under side of the head and throat, ferruginous salmon colour; under parts, light bluish; jaws furnished with teeth of a globular form; could perceive no second row on the palate. In other respects this reptile agreed with the description of *S. fasciatus*, the blue tailed skink of the New York Reptilia. The specimen appeared to have been killed some time, and had doubtless lost much of its true colour."

Snakes are unknown in the Bermudas, although we believe an instance has occurred of a dead specimen being found, which, however, was proved to have been transported in a truss of hay from New York. Long may their evergreen groves and grassy banks be free from these noxious reptiles!

PISCES.

The Bermudian waters teem with fish, and many West Indian species are to be found among them. Their presence here, some six or eight hundred miles from their native depths, may be owing to the influence of the Gulf Stream, which has been the means of introducing many tropical forms to the shores of these Islands.

It is much to be lamented that naturalists have hitherto left this interesting group of Islands unvisited, for most assuredly their marine zoology is well worth studying; situate as they are, at such a distance from any other land, in the midst of the ocean, the field they present to the ichthyologist is wide indeed, and well worthy of his investigation.

The different species of fish are known to the inhabitants merely by their local names, and a list we have by us, numbering 157 species, would edify our readers but little if we were to insert it. Such terms as "gogler," "cluck," "scuttle," "sailor's choice," and "slippery Dick," may be sufficient for the native fishermen, but we are afraid would not convey to others any clue as to their identity. Mr. Hurdis, with his usual kindness, has furnished us with the following notes upon a few species:—

COMMON SHARK (*Carcharias vulgaris*). The young of this shark are taken in abundance, and sold as food.

MACKEREL PORBEAGLE (*Lamna punctata*). This, the mackerel porbeagle of America, is commonly known by the popular name of the mackerel shark. One specimen, measuring $7\frac{8}{10}$ feet in length, was taken with hook and line in March, 1850. It is correctly figured in De Kay's New York Fauna.

SWORD FISH (*Xiphias gladius*). I have not met with this fish in the Bermudas, and the only opportunity I ever had of witnessing the effect of the amazing power of the *X. gladius* occurred to me on the 17th February, 1849, when the Bermudian schooner "Earl Dundonald" arrived in the port of Hamilton, having been pierced by one of these formidable fish off the coast of British Guiana, the particulars of which occurrence appeared in the Bermuda Royal Gazette of Feb. 27, 1849, in the following paragraph. "The schooner, 'Earl Dundonald,' belonging to D. R. Tucker, Esq., of Hamilton, sailed from Bermuda on the 11th Dec. last, under the charge of Capt. Wm. S. Doe.—Passengers: Robert A. Tucker, — Brown, of New York, and B. W. Watlington, Esqrs. When about 70 miles to windward of Demerara, and running down for that port, the vessel going six or seven knots, with a strong breeze, the man who was steering observed, for a short time, that the vessel broached to a little, and that she did not answer the helm readily. In the course, however, of two or three minutes the vessel was again under the control of the helm, and no further notice was taken of the occurrence. Subsequently a very small quantity of water was observed on the cabin floor, and it seemed to proceed from the locker, under and abaft of the cabin ladder; Capt. Doe was induced to examine the locker, and found the bill of a fish, about 9½ inches inside the ceiling on the larboard side, and so firmly fixed that it could not be moved. The bill is of a fish called by mariners the "bill fish.""

On boarding the vessel, in company with several gentlemen, to examine this great natural curiosity, the "sword" or bony spear of the X. gladius was found to have penetrated the larboard quarter of the vessel, within two or three feet of the stern post, passing through the copper sheathing—an outer planking of pitch pine 1¾ inch thick, the centre of a cedar timber 5 inches thick by 6 wide, and lastly through the ceiling of pitch pine 1½ inch in thickness. The weapon was nearly at a right angle with the keel, and pointed slightly upwards, the extremity projecting on the inside exactly ten inches beyond the woodwork. The point had evidently been broken off in the collision, leaving a jagged surface equal in size to the end of the little finger.

When the schooner had discharged her cargo, the "sword" was removed by cutting away the woodwork which surrounded it. It was dark brown in colour; measured 17¾ inches to the fractured base, and 5½ inches in circumference at the thickest part, and was split from the base to within about four inches of the smaller end. The upper surface was rough and granulated, resembling course sandpaper to the touch, and the sides exhibited a still rougher exterior. The under side was smooth and horn-like in appearance; the form not round, but oval; the upper and lower surfaces being of the greatest breadth. The smooth, or under side, was faced towards the bows of the vessel.

Yarrell mentions a dead sword fish found on the Essex coast in 1834, which was ten feet long, and of which the sword measured three feet. This rule of proportion would give five feet as the probable length of the "Earl Dundonald" specimen.

FLYING FISH (*Exocœtus* ———?). A very beautiful species, three inches in length, is the only one I ever examined. It was marked with indigo blue on the back, in the manner of the common mackarel, had two curious fan-like barbs suspended from the lower jaw, was barred with brown and yellow on the underside of the body, and the pectorals dark and crossed by two irregular bars of white. The ventral fins were nearly as large as the pectorals. This fish was washed on the rocks in rough weather. Another specimen *flew* on board the Salt Kettle ferry boat; Capt. Drummond considered it a new species.*

THE BONE FISH, OR GRUBBER. In January, 1849, a boat from Castle Harbour brought in a number of large fish I had never seen before. They were from ten to thirteen pounds weight each, and resembled the common mullet of the Islands in appearance, but with a longer and sharper head. The fishermen called them "rooters," and assured me they were rarely met with. A ship-master who was present described them as common in Jamaica, where they are termed "bone fish." Several hundreds were enclosed in the net, but not being sufficiently strong they dashed through it, and most of them escaped.

COMMON MULLET OF BERMUDA (*Mugil albula*). Very abundant at times, being found of large size in the shallow bays. Hamilton Harbour swarms with them, but from the

---

* On our voyage from Halifax to Bermuda, and within sight of the Islands, several flying fish rose from the water beneath the bows of the steamer, and after skimming about forty or fifty yards, would drop again. We did not observe one of them use the fins as wings, or *turn* in their ariel course. This species was about eight or ten inches in length, as near as we could judge from observation.

proximity of certain premises appertaining to the barracks, they are not exteemed in that particular locality.

THE SURGEON, or LANCET FISH (*Acanthurus phlebotamus*). Occasionally met with.

SPANISH MACKEREL, or HORSE MACKEREL (*Scomber colias.*) Very common. A commotion caused by some of these mackarel close to the rocks in front of Hamilton, induced me to approach the spot. They immediately retired, but not before the margin of the shore was strewed with "robins," and other smaller fish, which had escaped on shore. These I restored to their native element.

GOLDEN CARP (*Cyprinus auratus*). Abundant in ditches above the flow of the tide. It was introduced a few years since from Demerara.

ROBINS. The scientific name of these fish I am unacquainted with, but they are plump and handsome, about six inches in length.

HALF-BEAK, OR GAR FISH (*Hemirampus Braziliensis*). Common. I think this must be Cuvier's *Braziliensis*.

MARBLED ANGLER (*Antennarius marmoratus*). One which I examined was taken in a fish pot in Hamilton Harbour. Another was captured among floating seaweed (*Fucus natans*) by the crew of a vessel homeward bound from Turks Island. This I have in spirits; length $2\frac{6}{10}$ inches.

SPOTTED THREAD HERRING (*Chatœssus signifer*). Of this fish I have seen the young only, which were about the length of a man's finger.

COW PILOT (*Chætodon* ——— ?). Commonly known in the Bermudas as the "pilot fish," but proved by Capt. Drummond to be a species of Chætodon.

EELS (*Anguilla*). I know of one species only, which is

common in all the ditches or dykes of the Pembroke marshes. It resembles the *Anguilla vulgaris* of Europe, sometimes attains two or three pounds in weight, and is very destructive to young ducklings. The 'Mudians are prejudiced against this fish, and will not eat it.

BANDED PIPE FISH (*Syngnathus fasciatus*, De Kay). Length six inches. One specimen only obtained by me.

PORCUPINE FISH, or SEA HEDGEHOG (*Tetraodon* ———?). Very common.

SEA HORSE (*Hippocampus brevirostris*). Not uncommon.

Oct. 31, 1851. A fisherman drawing his nets for turtle in the harbour of Hamilton captured a curious fish, which he requested me to look at. I found it attached to a small line, swimming about the fisherman's boat. When taken out of the water it uttered a grunting noise; length, twelve inches; extent, fifteen. Colour of the upper parts, greenish-brown; membrane of the *wings* black, spotted with blue towards the extremity. Two long powerful spines were recumbent on the back part of the head, and numerous small spines presented themselves on the posterior part of the body towards the tail. Could find no description of this fish in any work at the public library, except a brief account in Adam's Encyclopedia, from which I am inclined to think our specimen to be the common Dactylopterus, or Flying Gurnard of the Mediterranean (*Trigla volitans of Linnæus*), which is described as common in that sea, though entirely unknown along the oceanic coast of Europe; also as being found on all the central and southern shores of America, even as far north as Newfoundland. This distribution of Dactylopterus volitans is supposed to be aided by the waters of the Gulf Stream.

The fish appears to be unknown to the Bermuda fishermen.

A small fish known by the common name of "Anchovy," but which I believe to be the Sardine of commerce, (*Clupea sardina,*) frequents the shallow creeks and harbours of the Bermudas in immense shoals, in February and March. It averages about five inches in length.

A Bream (*Abramis?*) is also a very common fish, large shoals of which appear in the shallow waters. These are frequently enclosed in nets, and kept as a reserve from whence to supply the daily demand.

The Sennet is likewise a common fish in the waters of Bermuda, and sells freely in the market.

MOLLUSCA.

The reefs which extend in every direction, save the south and east for many miles around the Bermudas, present ample scope for the labours of the conchologist, as do likewise the sheltered harbours, sounds, and inlets, for which this group of islands is so remarkable. Although several individuals, at various times, have made small collections of shells while staying in the islands, we have heard of no collector, of scientific attainments, having visited their shores for the purpose of thoroughly investigating their marine zoology. It is, therefore, to be hoped that ere long, some enthusiastic naturalist may be induced to visit this much-neglected group, and let the dredge scrape the bottom of those hitherto almost totally unvisited depths, and so bring to light many rarities, which we doubt not are now lying hid amid the coral groves, and the chalky mud that surrounds them. The brief list of marine and land shells, which we here present to our readers, is made up, in a great measure, from a list which appeared in the "Bermuda Almanac," for 1852, of a collection made in the Bermudas, by Dr. Temple Prince, of New York; but we have been enabled to make some additions to that list, from specimens we collected during our visit. We are also indebted for three extra species to the kindness of Mr. Cuming, who, with his usual courtesy and love for his favourite study, greatly assisted us in naming our specimens.

The land shells collected by Dr. Temple Prince, are as follows: *Helix palludosa*, *H. ptychoides*, *H. selenina*, *H. Bermudensis*, *H. Sancta Georgiensis* (n. s.), *H. Somersetii*

(n. s.), *Bulimus ventrosus, B. Bermudensis* (n. s.), *B. Sandysii* (n. s.), *Pupa Bermudensis* (n. s.), *Helicina variabilis, Succinea Bermudensis,* (n. s.)

To these we add from our own collection, *Helix microdonta,* Desh.

The Marine species collected by the same gentleman comprise: *Scalaria tenuis? Vermetus Knorrii, Cerithium Greenii, C. litratum, C. Eriense, C. Bermudensis* (n. s.), *Littorina albescens, L. muricata, Columbella cribraria, C. mercatoria, Buccinum ambiguum, Purpura fasciata, P. deltoida, Bulla occidentalis, Conus mus, C. daucus, Oliva nivea, Nerita versicolor, N. tesselata, Neritina viridis, Turbo pica, Trochus modulus, Chiton squamosus, Siphonaria picta, Lutraria dilatata, Marginella avena, Terebra hastata, Dolium perdix, Fissurella græca, Truncatella aurea* (n. s.), *Auricula flava, Cypræa rotunda, Lythodomus dactylus, Spirula Peronii, Pecten ziczac, Arca Noæ, Mytilus exustus, Perna ephippium, Lima scabra, Tellina lævigata, Lucina tigrina, L. pecten, L. squamosa,* and *Cythærea Bermudensis,* (n. s.)

To these we can also add, *Littorina Mauritiana, L. ziczac, L. dilatata, Nerita peloronta, Modiola tulipa, Cardium serratum, Tellina Gouldi* (of Hanley), *Arca lactea, Purpura undata, Oliva bullula, Semele subtunicata, Lucoma* ——*?*, *Melanigrena placunoides,* (called the "Bermuda Oyster," by the natives,) *Rissoina pulchra,* and *Pholas striata.*

The three species alluded to as being in Mr. Cuming's collection, are *Spondylus ramosus, S. longitudinalis,* and *S. digitatus.*

Mr. Hurdis informs us that there is also a large species of *Pinna.*

INSECTA.

THE Entomology of the Bermudas does not present that ample field we could desire; the limited area of those isles, their isolated character, and other causes, alluded to in these pages, sufficiently explain why the forms of insect life should be confined to few genera. We believe that most of the insects found in the Bermudas, are also to be met with in the nearest portion of the American continent,—North Carolina, distant about six hundred nautical miles;—thus leading us to infer that each species, at some former period of time, and by methods not clearly understood, has been transported over sea from the Virginian coast to its present habitat. In like manner, the few native insects common to the West Indies, are supposed to have been introduced from thence in the packages of shrubs, plants, or fruit, occasionally imported from that quarter.

COLEOPTERA (BEETLES).

The little modest colored *Cicindela tortuosa* is found in some abundance in the Bermudas, where it is common on the rocky point of Harris's Bay, in the months of September and October. Like its congeners of other lands, this insect is ever on the alert, and ready to spring on the approach of the collector. It is the only species of the family *Cicindelidæ* we met with on the islands.

By far the most common beetle, and the best known to the inhabitants, is the "hard back" (*Ligyrus juvencus*). The strength of this insect is amazing, considering its small size; and if a glass tumbler or a candlestick be placed upon its back, it will readily disengage itself. It appears to be quite nocturnal in its habits, flying in numbers into the rooms when the candles are lighted; and if taken in the fingers, considerable pressure is necessary to prevent its escape: during its struggle for liberty, it utters a short squeaking noise. A single specimen of a species of *Heliopates*, we found beneath a stone on Cooper's Island. *Diaprepes affinis* is rather common, usually frequenting the wild sage bush when in bloom. It is a pretty insect, and appears to be covered with a bluish-green powder, which readily rubs off when the insect is pressed between the fingers. An *Elater*, of a light-brown colour, frequents the trunks of the cedar trees, but we are not certain as to the species. Mr. Hurdis mentions a small species of grain-eating beetle, probably a species of *Bruchus*, which he states to be extremely destructive in the stores of Indian corn, meal, biscuits, rice, &c.

Doubtless a few other species of coleoptera might be added to the list, if proper search were made in the spring of the year; but, unfortunately, our visit being in autumn, all our efforts were unavailing to increase our collection, although we worked hard almost daily during our stay.

ORTHOPTERA (CRICKETS, COCKROACHES, LOCUSTS, &c.)

The common cockroach of the Bermudas is *Blatta Americana*. It is extremely abundant during the hot summer months, particularly in old houses surrounded by trees. Far different from our English species, which appears to have a preference for "life below stairs," the *B. Americana*, strong and swift of wing, flies in through the drawing-room windows, and, buzzing about the heads of those assembled at the social tea-table, pops down upon any substance within its reach; now on a piece of cake,—now on the butter,—off again to the milk jug,—then gnawing away at the candle, its long feelers occasionally getting a reminder from the flame, until at last defeated of its purpose, it flies off through the open window, to become troublesome to the inhabitants of some other abode. Nothing apparently comes amiss to these insects as food, although they generally show a partiality for substances of a greasy nature. Among other accomplishments, they will drink black ink with impunity; lay a hat by for a few days, and on taking it down from its place, you will generally find the leather band inside gnawed all round by these mischievous creatures; a pair of dress boots will share a similar fate, the enamel being eaten off in several places. The 'Mudians generally use bags (to preserve extra boots and shoes in), well tied at the mouths, to keep out these indefatigable destructives. It is rather a handsome insect, of a chesnut colour.

All kinds of poultry feed greedily upon the cockroach; tame ducks spending entire moonlight nights in their capture.

Another species of cockroach, far less numerous, and rarely seen, except in cellars, and other dark places, is *B. Maderensia.* It is known on the islands under the name of "*Knocker*," from a habit it has of making a noise like a person gently tapping a box, or skirting board. It is much larger than the former species, and the wings are of a whitey-brown colour. A tub of Indian corn is its favourite resort.

Under the stones, in the cedar groves, we found vast numbers of apparently two distinct kinds of wingless cockroach, but whether they were merely the young of *B. Americana* and *B. Maderensia*, or a totally different species, we have not been able to ascertain. The larger measured about ¾-inch in length, having the whole body of a dark chesnut brown colour; the other kind somewhat smaller in size, having the body banded with yellow and dark chesnut brown stripes.

After a day or two of cold weather in October, we observed, when turning over stones in search of specimens, that these cockroaches were benumbed and perfectly helpless. In similar situations, we almost invariably found a large species of cricket, in its various stages of growth, from youth to maturity; the latter stage having the wings fully developed, while in the former, merely small sprouts could be discerned. It was from these facts we concluded (though, perhaps, incorrectly), that the wingless cockroaches were neither more nor less than the young of *B. Americana* and *B. Maderensia*.

On the sloping banks of the well-known sand hills, in Paget's parish, we first became acquainted with a small, yellowish-brown coloured grasshopper. It reminded us greatly in its habits, of our common green-coloured chirping friend of Old England, and was just as difficult to capture. The insect is common on the open tracts, particularly where the sandy waste is relieved by tufts of grass.

A species of locust, which we take to be identical with the common locust of the American continent, is found in similar situations, although not so numerously. On the parade ground, in rear of the barracks, at St. George's, this insect is in some abundance during the month of September. A ludicrous account of the capture of a specimen, now before us, we may be permitted to relate:—While walking one hot day in the vicinity of the above-named barracks, with our kind and lamented friend, the late Colonel Oakley (56th Regt.), on the look-out for insects, a very fine specimen of this locust sprung up before us; we chased it for awhile unavailingly, but determined not to be baulked of our prey; the Colonel then joined in the pursuit, and after a sharp and hot chase, our gallant companion finally bagged his game right before a sentry box; the sentry, as in duty bound, standing with arms presented, in the presence of a field officer, who was, however, in a rather undignified position to receive a salute; we had gained our prize, however, and had a hearty laugh, in which we fancy the sentry could scarcely help joining.

NEUROPTERA (Dragon Flies, &c.)

There are three, if not four, species of true dragon-flies (*Libellulidæ*), frequenting the marshes and ponds in the Bermudas; but as our specimens were unfortunately destroyed by the ants and cockroaches, we failed in identifying them.

It is by no means a pleasant pursuit, to frequent the marshy tracts in search of insects, at mid-day; for the decayed vegetable matter, while stewing under the intense heat of an almost tropical sun, emits fetid exhalations, hardly endurable, at least by an English constitution.

STEGOPTERA (Ant-lions, May Flies, &c.)

A species of Ant Lion (*Myrmeleon*), is common on the islands, which, from their sandy character, are well suited to the habits of this singular insect.

HYMENOPTERA (Bees, Wasps, &c.)

One of the most common insects of this order, is *Polistes pallipes*, commonly known to the inhabitants as the "Bermuda Wasp." The nest, or, more properly speaking, comb, of this species, is usually fastened to the branches of the wild sage (*Lantana salvifolia*), and is about the size of a man's hand; flat, with its upper surface shiny and apparently waterproof, having the mouths of the cells reversed, or opening beneath, of which we counted some two hundred and eighty in a specimen we brought home and presented to the British Museum. When constructing their combs in the cedar groves and cultivated parts, they appear to choose the proximity of a wall, in preference to the open, perhaps for shelter from the winds. Mr. Hurdis informs us that their combs abound in the marshes, where they hang suspended from the stems of the tall sedges. It is a cowardly insect, and is not half the plucky fellow our English wasp is, who always guards his nest with vigilance, and attacks the intruder with the greatest fury; but our friend of Bermuda will not only let you come close to his comb, but will positively allow you to cut off the branch whereon his abode rests, without any molestation on his part. It is prettily marked, of a lightish brown colour, striped with bands, brown and buff, across the abdomen.

A handsome, although rare insect, is *Pelopaus ceruleus*; known on the islands as the "blue wasp." We only saw one specimen during our stay, and this we fortunately captured, in the garden at Hermitage. It has the thorax and abdomen of a bright metallic dark-blue, and the wings tinged of a like colour. Another species, *P. flavipes*, is rather common, having its body prettily marked with bands of black and yellow. For our own satisfaction, and to clear up a doubt, we allowed one of these insects to sting us, and found the pain somewhat similar to that produced by the venom of our English wasp. The flowers of the fennel plant seem to attract this, as well as the former species. Mr. Hurdis states that it is known on the islands as the "mason wasp," and that "it builds its clay cell upon ceilings of verandahs, eaves of houses, and about windows." He remarks, "it is a fine insect (somewhat larger than the Bermuda wasp), and may be observed throughout the summer, carrying, within its legs, a round ball of moist, sandy clay; with these balls it moulds its cell with much industry, making a peculiar buzzing noise while so engaged. When completed, the structure resembles an oval lump of mortar. Two or three cells are sometimes united. On examination, the interior of each will be found in the form of an oblong apartment, neatly rounded, and of sufficient size to hold the parent insect; into this the mason wasp introduces as many spiders,—alive, though apparently half dead,—as the aperture will hold; and having deposited its egg among them, closes up the entrance with clay in a very neat manner, leaving the future care of its progeny to Nature. The larva, after consuming the unhappy spiders, assumes the chrysalis state, and comes forth a perfect insect."

The "honey bee" of the Bermudas is very common; and, apparently holding to the old maxim, "improves each shining hour," as do its congeners in more temperate climes. It is smaller than the common honey bee of England; and, from comparison with specimens in the British Museum, we are inclined to believe it to be *Apis caffra*. Mr. Hurdis supplies us with the following note:—"The honey bee is common in the Bermudas, and may be seen in many of the cottage gardens; not in the straw-made hives of England, but in square wooden boxes, ranged upon a wooden bench. They make honey of good quality, which sells at the rate of two shillings per common quart bottle.

"The bees near Hamilton draw largely on the sugar warehouses in the town. The windows being constantly open, they pass between the iron bars, and, finding every hogshead and barrel bored with holes at each end, they enter the packages by hundreds, and carry away no small quantum of sugar in the year.

"I was once at a large picnic party on Port's Island, and, while lolling on the grass near the tables at which the party had feasted, some one pointed out a magnificent comb of the honey bee, in a miniature cavern, or recess, in the rocky bank close by. It was about two feet in depth, and was suspended from the upper surface of the rock, so that you could lie on the grass and contemplate the busy scene within three feet of you. It was a beautiful sight, and proves that the bee does not require to be boxed up in darkness, in a climate like that of the Bermudas. I cannot pretend to estimate the weight of that comb, though I am quite sure an ordinary beehive would not have contained it."

Mr. R. S. Wood informs us that the comb of the honey bee is also found in the caves at Walsingham.

A singularly formed insect (*Evania appendigaster*), which has its small posterior body, joined by a lengthened and slender stem to the thorax, is not uncommon; but of its habits we know nothing, save that it has a peculiar way of vibrating the aforesaid appendage up and down in a rapid manner.

The Common Ant (*Formica*), of the Bermudas swarms in countless myriads throughout the hot season, and makes sad havoc among the naturalist's specimens, during the drying process; but should a univalve, or crustacean, have to be cleared of its fleshy inhabitant, certain are we that a more sure method could not be resorted to, than to place the specimen within reach of these never weary scavengers; but a short time would elapse ere every particle would be consumed, save and except the shelly covering, which, by this simple process, would be prepared for the cabinet.

The following interesting note, by our friend, Mr. Hurdis, more clearly illustrates the habits of this insect:—

"Every housekeeper, gardener, and all who deal in articles of food, are keenly alive to the destructive habits of the ant, which infests the Bermudas in legions during the greater portion of the year. Nothing appears to escape their active search; and whether it be meat, milk, sugar, honey, cake, or fruit, the ingenuity of the owner has to be exercised, in order to save the same from utter destruction.

"There are two species of ants in the Bermudas, one of which is about the size of the common ant of England, and is supposed to be an importation from the West Indies: the other is a much smaller insect, which I have observed only among the islands in the Sound.

"The ant appears to entertain a natural repugnance to common whale oil, for which reason it is commonly used by the native and other inhabitants as a protection against its annoying depredations. Store-room tables have their legs placed in tin or leaden cups, partly filled with this oil. Shelves, made to hang from the ceiling, have their iron supports passing through tin funnels of the same; and meat hooks are guarded in a similar manner. It is only by these means that any article of food can be considered safe from these marauders. The ant is also very destructive to the domestic rabbit, to poultry, and young pigeons, to caged birds, and to all sick animals; and man himself, when in a helpless state, is sometimes attacked by it.

"During the heat of summer, millions of these insects make their appearance upon every road and pathway, and sometimes invade the dwellings of men in such multitudes, as to become an intolerable nuisance. Every tree, and almost every bush then teems with its black columns, ascending and descending in the great occupation of obtaining food.

"Finding the ants, one morning, disposed to attack a bottle of honey (a common wine bottle), I placed the same in a soup plate upon the sideboard, carefully filling the plate with water as a protection. On returning to the room a short time afterwards, I found the bottle swarming with ants, and on a closer inspection was greatly surprised to find a column of those insects passing and repassing on the surface of the water, between the rim of the plate and the bottle of honey. This they appeared to do with perfect ease, merely wetting their feet in the operation; in other words, absolutely walking on the water.

"There is another peculiarity in the habits of this insect, which deserves to be mentioned. If a couple of snipe have been killed, and are destined as a present to some friend, they will be suspended by a single thread from the upper part of some cool, open window for the night, to save them from the ants; and yet, notwithstanding this precaution, the birds will be found in the morning covered with ants, while others continue to descend the long and slender passage of communication. Is not this property of discovering food to be ascribed to the power of scent?"

A species of *Myrmicidæ* we found under a stone, in the grounds at Hermitage, in company with another small red ant.

LEPIDOPTERA (Butterflies and Moths).

Of the few species of diurnal lepidoptera frequenting the Islands, the Archippus Butterfly (*Danais archippus*), ranks first for size and beauty.

This fine species, which is so frequently seen in the southern states of America, is also common in the Bermudas, and generally to be found where a patch of ipecacuanha is in bloom. Mr. Hurdis' note book has the following:— "Dec. 18. The beautiful caterpillar of *D. archippus* is still common. It is two inches in length, and elegantly encircled from one extremity to the other with bands of black, white, and yellow. The chrysalis is of a delicate pale green colour, semicircled with a line of black and gold, and dotted in various parts with bright golden spots. It makes no cocoon, but suspends itself by a black ligament from the tail end, to stone walls and railings, and continues in the pupa state exactly twenty days. The favourite food of the caterpillar of this species is the curascoa swallow-wort, or bastard ipecacuanha (*Asclepias Curassavica*), which grows wild in pastures, gardens, and roadsides, and blossoms during every month of the year. The perfect insect is abundant throughout the year."

Another member of this family, *D. Berenice*, is of rare occurrence in the Bermudas; Mr. Hurdis having only seen some three or four specimens during his long residence on the Islands—one of which, captured by Mr. Hunt Marriott, of H.M. Customs, on the 5th July, 1850, was of a drab or cream colour.

The Red Admiral (*Vanessa atalanta*) is not uncommon, and is met with from April to November. It frequents gardens and road sides, the larvæ feeding upon the common stinging nettle, which grows freely in many places. It is probably the American species, to which Sir Charles Lyell, in his "Second Visit to the United States of America," alludes in the following words:—"In one of the cabinets of Ohio insects, I saw (at Cincinnati) specimens of that common English butterfly *Vanessa atalanta*, or red admirable, which I had observed in the winter flying about in the woods of Alabama. I could not distinguish it from the European species, yet Mr. Doubleday, the entomologist of the British Museum, at once recognised all I shewed him as American specimens; for there is a minute but constant difference, first pointed out by Mr. Stephens, in the markings of the beautifully coloured anterior wing."

The Painted Lady (*Cynthia cardui*). We have observed this butterfly on Gibb's Hill, near to the lighthouse, in the early part of November. Mr. Hurdis also states that it is common on the North-side hills, from August to November; and that in the year 1852 it was much more abundant than he ever remembered to have seen it before.

The Camberwell lBeauty (*Vanessa Antiopa*). We are induced to insert *this species* in the list from the following observations of Mr. Hurdis:—"A few years ago I observed near the gaol of Hamilton a brownish butterfly, with a broad white or yellow margin to the wings. It was precisely similar to V. antiopa, the Camberwell Beauty of Europe, and the species being common in North America, I have always considered it to be a stray specimen imported in the pupa state with shrubs or flowers from that part of the world."

V. cœnia is exceedingly abundant in the Bermudas, and is known to the inhabitants by the name of "musk butterfly" —probably from its apparent fondness for the melon patches, where it is generally to be found basking on the leaves. It is a most pugnacious little creature, and appears to love a quarrel, for you may see three or four of them ascending in the air and buffeting each other, now rising, now falling, unremittingly continuing their aerial warfare, till lost to sight behind some neighbouring cedar. This species is widely distributed, and is to be found in the tropical districts of both hemispheres. The specimens from the Islands of the Pacific are apparently more highly coloured than the 'Mudian ones. This insect is often named *V. orythia* in cabinets.

Terias lisa, of Boisduval, was first observed by Mr. Hurdis on 10th Oct. 1847, on which day it suddenly appeared in great abundance, hundreds being seen in every direction. Previous to that occasion he had never met with this butterfly, although accustomed to keep a good look out. They all disappeared, however, in the course of a few days. In other seasons he considered them comparatively scarce, appearing for a few days only in September, October, or November.

During our rambles through the cedar groves in the autumn of 1854, we observed on two separate occasions a good sized yellow butterfly, but unfortunately failed to capture a specimen, as its flight was very rapid, and it is difficult to chase an insect through the cedars and dwarf scrub. Mr. Hurdis evidently refers to this unknown rarity in the following note:—"Sept. 23, 1853. A beautiful butterfly of a delicate brimstone yellow, tinged with a greenish hue, and fully as large as the sulphur butterfly of England, appeared on this day, flitting with wonderful activity about a patch of sweet potatoes in my garden. It baffled all my attempts to take it. This is the only instance of my meeting with this butterfly, the name of which I am unacquainted with. On the 8th and 9th of the following month it was seen in the same garden by my son, who followed it across the adjoining properties of Mr. Kennedy and Lady Burnaby, without being able to capture it. He described it as being the colour of 'mustard,' rapid in flight, and seldom settling."

Mr. R. S. Wood, of Walsingham, informs us that some time ago he captured specimens of this butterfly.

Our list of nocturnal lepidoptera is confined to three species, and although we tried the usual method, of going out with a lamp at night, and resorting to other expedients, we failed to add to our collection. The largest and handsomest species known on the Islands is *Sphinx cingulata*, which is not uncommon, frequenting the flowerets of the night-blooming shrubs. A shrub of this description, standing against the house of Chief Justice Darrell, at Cavendish, was a favorite resort of this species. Cats are very fond of catching this insect, whether for food or mere pastime we cannot say; however, we never saw one eaten by them. The caterpillar of this sphinx grows to a large size, and feeds on the green shoots of the sweet potatoe. The Pink-underwing moth is found in abundance from the middle of August to November, and a small brown species of *Tinea*, the only specimen obtained, we captured at Hermitage.

HEMIPTERA (Cicadæ, Bugs, &c.)

One of the *Cicadæ* (*Fidicina tibicen*), a very noisy individual, very appropriately named "scissor grinder," or "singer," by the inhabitants, may certainly be heard, if not seen, during the hot weather. All day long, while the sun shines, this insect keeps up a continual buzzing noise, which is heard to perfection (if we may be allowed the expression) when caught in the folds of the entomological net. It is a very quick sighted insect, and is difficult to capture; ceasing its note on the approach of the collector, it remains perfectly motionless until the net is drawn towards it, when off it starts with a swift jerk and a loud buzz of apparent derision.

On the approach of cold weather, or indeed even in summer, when the sun for a time ceases to pour down its scorching rays, the "scissor grinder" issues merely a low subdued note, but no sooner does the mid-day orb escape the shadowing cloud and once more emit its gladdening beams, than our friend the cicada raises his tone in apparent ecstacy; and when all else in nature seems wrapt as in a midnight slumber, our ever wary, yet never weary, "singer" trills his measure, cheering the glades and evergreen groves of these fairy Isles with his merry, albeit vociferous and thrilling clamour.

We believe this to be the only species found on the Islands.

A goodly sized flat green bug (*Rhaphigaster prasinus*) is very common during the hot months of summer, and is a source of great annoyance to those affected with a delicate sense of smell, for in common with others of its race, it emits a peculiar odour when touched. It usually flies in at the open windows during dinner and darts right at the candles, which, being surrounded by glass shades, cause the intruder to fall stunned on the table cloth, from which it is speedily removed by the attendants. We first became acquainted with the species at the Artillery Mess at St. George's, where a specimen popped down close to us at table, and not having seen one previously, we were on the point of securing him as "a rare specimen" for our collection, when a sharp-eyed mess waiter *anti-entomologically* inclined, removed him from the table. A very small species (*R. cydnus*) is by no means common.

DIPTERA (Flies, Gnats, &c.)

As in other parts of the globe, and equally common in the marshy tracts, the well-known and troublesome mosquito (*Culex*) finds a home even in the distant and small ocean-surrounded islets of Bermuda. Its bite, which is just as sharp in those Islands as elsewhere, is only felt to perfection when looking out for ducks at the edge of a marsh, at dawn of day; we would defy the most ardent sportsman to act up to the well-known instructions of laying "still as a mouse," when some dozens of these blood-thirsty fiends are stabbing him in every unprotected part.

Mr. Hurdis considers there are two, if not three, distinct species, for he thus writes:—"The plumed mosquito, I am quite positive, is distinct from the common gentleman, whose acquaintance is but too familiar to me. The former is magnificently plumed, and not quite so large an insect as the common species; besides which, the larvæ are totally different; the one being a small blood-coloured worm, and the other the restless wig-wag so often seen in our tumblers.

"The larger mosquito, which is found in swarms—distinct swarms, I mean—in the marshes, is certainly different from the two species I have mentioned, being of a larger size, with the body much distended with something very like mud; for in killing one on the person it leaves a muddy spot. A swarm of these mosquitos may be seen and heard at some distance as they dance up and down over the low cedars."

Three species of ***Muscidæ*** have been noticed by Mr. Hurdis:—A house-fly, very common and troublesome in some seasons, which is closely allied to (if, indeed, it be not identical with) *M. domestica* of Europe; a grey coloured fly, which produces its larvæ alive; and the common black one, which lays eggs. The two latter are both blow-flies, though by no means abundant. A very beautiful blue-bottle fly is also frequently met with; as also a light yellow-coloured species of cow-dung fly (*Scatophaga*), which frequents the droppings of cattle, very similar in appearance to *S. merdaria* of Britain. We also observed a species of *Tabanus*, of rather rare occurrence. A green-headed gad-fly is not uncommon; Mr. Hurdis takes it to be *Gastrus equi.*

APTERA (FLEAS, SPRING-TAILS, CENTIPEDES, ETC.).

The Common Flea, (*Pulex irritans*), is extremely abundant in the spring-time of the year, appearing as the 'Mudians say, "at the same time as the whales on the coast." It is doubtless the same species which distresses the dogs so much. The Chigre, (*P. penetrans*), so well-known in the West Indies, is also common in Bermuda, and many are the tales related of its dangerous attacks.

Centipedes (*Scolopendræ*) are fortunately not common in the Bermudas, and those that do exist frequent particular spots where they are seldom disturbed, such as lumber heaps, roots of old cedar trees, and other like places. Specimens have been taken of great size; one captured at St. George's was kindly presented to us by Mr. Attwood, and measured about six inches in length, including antenæ and hind legs; but this is small in comparison to some, which, from their venomous character and formidable appearance, are much dreaded by the inhabitants.

Small species of the Millipedes, (*Julus*,) are frequently met with, and are locally termed "forty claws."

ARACHNIDA (Spiders, etc.)

Five species of spider came under our notice, of which the "silk spider" (*Epeira clavipes*) is the best known and most attractive. This wonderful web-architect is extremely common, and its strong and extensive web spreads from tree to tree in all parts of the islands. It appears to have been noticed by travellers as early as the seventeenth century, at which period some remarks of its habits appeared in the "Philosophical Transactions," stating that "they spin their webs betwixt trees standing seven or eight fathoms asunder; and they do their work by spinning their web into the air when the wind carries it from tree to tree. The web when finished will snare a bird as big as a thrush." Now we think this statement to be an exaggeration; if not, the habits of this insect must have changed since those days, for we never saw a web stretching anything like seven fathoms (42 feet) in breadth; and think we should be much nearer the mark if we were to place the greatest breadth at twelve to fifteen feet. Again, a bird the size of a thrush would easily break through the strongest web, and we believe it is only the small and barely full-fledged young of the white-eyed greenlet (*Vireo noveboracensis*) that has been known to become entangled in the meshes of this web.

Being particularly anxious to test the strength of the silk, we one day caught an Epeira, and taking hold of the end of the silken thread hanging from its body, began to wind upon a piece of paper, and succeeded in obtaining a few yards of beautiful light yellow-coloured silk,—and this single thread was so strong that we could pull it slightly without breaking it. We were informed that the 'Mudian ladies made use of it for sewing purposes. The insect is of a large size, having, with fore and hind feet, a span of four to five inches. It is prettily coloured; body yellow, with white dots; head, dark brown. How they find food, and where, in the hot weather, we are at a loss to guess, for their webs contain only a few scattered remains of cicadæs and "hard backs," barely sufficient we should think, to support the plump and rotund figure of Epeira clavipes. Towards the close of the month of October, when the fierce equinoctial gales, accompanied with heavy rains, put an end to the bright hot sunny days of summer, the "silk spider" and his web alike disappear together; and, instinctively taught that fruitless it would be to endeavour to repair damages, and withstand the fury of the elements, he adopts the wiser course of ensconsing his form in some secure retreat, from which he emerges in the first bright days of spring, again to spin his ample web, and spread its meshes across the open glades, and from branch to branch of the sweet-scented cedars.

The large house-spider of Bermuda is quite as large as the former species, but different in shape and colour. In the hot months, before a gale commences, this spider makes its way into the houses, sometimes carrying under its body the bag containing the young, and takes up a position in some snug nook or corner until the storm be past. We took a bag from one of these insects for a specimen; it is exactly the size of a sixpence, flattened in shape, and rounded at the edges. Mr. Hurdis furnishes the following note concerning this species. "April 19th, 1855. Measured the span of the common large brown spider on the parlour wall, by rule and compass, and found it to be four and a half inches. This spider had inhabited the same room for many weeks."

A species with very minute body and amazing long legs of most elegant and slender proportions, inhabited an out-house at Hermitage, where, in the corners of the walls, were several separate webs of small size, each occupied by one of these insects, which, on our approaching close, commenced vibrating its web so quickly as to make the spider invisible to the eye. This species is known under the name of "Ringer" on the Islands.

Two or three other species, about the size of our common English Spider, were common under the stones in the cedar groves, and a very pretty kind about the same size, and of a handsome grey colour, we observed once at the edge of the marsh below Verdmont.

We may close our list of Arachnidans with a species of Tick (*Ixodida*), which attacks cattle and dogs.

In our brief list of insects we must not omit to notice the occurrence of the Firefly (*Lampyris*) in Bermuda, on the authority of our friend Mr. Hurdis, who kindly sends us the following from his note-book. "August 14th, 1850. Was very much surprised this evening (which is calm and dark) to see a bright luminous spot moving through the atmosphere in a somewhat undulating course towards the open windows of my residence, on approaching which it rose towards the eaves and disappeared. I immediately exclaimed, 'a firefly,' and ran out of the cottage with the view of watching its movements; it was not, however, to be seen."

"The firefly of North America is familiar to me, and I should certainly say *is much smaller* than the brilliant insect so accidentally observed to-day, the phosphoric light of which appeared to be delicately tinged with green. This is the first instance of a firefly being seen in the Bermudas that I am aware of. The question will naturally arise, as to whence, and in what manner it could have been introduced. My own humble opinion leans to the probability of its having been brought here from the West Indies, in the many boxes of plants occasionally landed at these Islands.

"Mr. Trimmingham, of Hungry Bay, subsequently assured me that he had more than once observed the firefly in Bermuda, and that its occurrence was beyond all doubt."

CRUSTACEA.

Common edible Crab, of the United States, (*Lupa dicantha*).

Spider Crab, or Sea Spider, (*Lichinia canaliculata*).

Land Crab (*Gecarcinus ruricola*). This is of a good size, and nocturnal in its habits. It is the colour of unbaked clay.

Pericera cornuta. Taken from a fish-pot, at Hamilton, 11th April, 1850. Length 7·5 inches; breadth, from spine to spine, 4·8 inches.

Soldier, or Hermit Crab (*Pagurus* ——— ?). Not uncommon.

Long-tailed Crab, Stump, or French Lobster (*Scyllarus equinoxialis.*)

Cray Fish (*Palinurus* ——— ?) called in the Islands "lobster." It is of large size, and sufficiently abundant.

Sand Bug (*Huppa* ——— ?) Buries in the sands.

Common Prawn (*Palæmon serratus*). Attains to a good size, viz:—5⅓ inches in length, or including the antennæ 11¼ inches. Not uncommon, if sought after.

Common Shrimp (*P. vulgaris*). Common in the rock pools at low water. Mr. Hunt Marriott sent Mr. Hurdis a plateful of these shrimps which he had taken with a landing-net in Hungry Bay.

"There are other species of crabs," says Mr. Hurdis, "the names of which are unknown to me, viz:—one which climbs the mangrove trees, very rosy on the under side; one which the fishermen call the 'coral crab,' quite as large as any crab found on the Islands, of a greenish brown colour on the upper parts, with very long claws, the outer joints of which are margined with roseate or carmine; back, claws, and legs beautifully furnished with numerous short white spines, thick and strong; the outer margin of the shell, also armed with these spines, four of which project forwards over the mouth. It differs much from *Pericera cornuta*, of the West Indies."

The vignette below represents Gibbet Island, Gibbon's Bay, as seen from the North-road above the Flatts. It is taken from a sketch kindly forwarded by Miss L. L. Deudney.

BRACKISH POND CHURCH.

BOTANY.

AT a short distance from the land the Bermudas appear to be clothed in some parts with verdure, and in others by a thick growth of evergreen trees. On a nearer approach, the former will be found to consist of the trailing crab grass and common sage bush, and the latter of one species of cedar, not indigenous to the group, as its name would apply, but common also to that portion of the American continent, lying in the same latitude.

From accounts given by early navigators, these islands appear to have been well wooded with this hardy and densely growing tree, so useful in the present day for building purposes, for fuel, and for the shelter it affords from the violent gales of wind which so frequently occur during the winter cultivation of the soil. It is also admirably adapted, by its resinous, unshrinking, and durable qualities, for building vessels of moderate tonnage.

In the following list, six species of Citrus are mentioned as growing in the Bermudas; a seventh, called the "Grape Fruit,"—almost twice the size of an orange—is also found there, the specific name of which we are unacquainted with. They are all cultivated in gardens, with the exception of the Lemon, which grows spontaneously everywhere, adorning the roads and hill sides with the abundance of its golden yellow fruit, of which thousands of bushels annually fall only to rot upon the ground.

Delicious beyond description is the perfume emitted from the expanded blossoms of these fruit-bearing trees, and more particularly of a calm evening, after a copious fall of rain, when the sun reappearing in subdued brightness and splendour, gilds each cedar-crowned hill and white-washed cot with its fading beams, the powerful scent of the citron tribe, mingled with that of the cedar, is exhaled in such copious quantity, as forcibly to impress the imagination with a realization of those fairy lands of ancient fable, where gorgeous palaces, inhabited by rich and happy princes, were fanned each live-long day by balmy breezes, heavy laden with odorous incense.

The following brief account of the trees, shrubs, &c., now growing on the Islands, emanates, with but few exceptions, from the pen of Mr. Hurdis. It is not intended by any means to represent a complete list, but will be found to contain the names of those species best known to the inhabitants.

Orange (*Citrus aurantium*). Cultivated in gardens.

Common Citron (*C. tuberosa*). Cultivated in gardens.

Seville, or Wild Orange, (*Aurautum acre*, of Miller.) Grows wild.

Sour Lemon, or Lime, (*C. acris*). Grows wild.

Common Lemon, (*C. limon*). Grows wild.

A few years ago, about 1854, the orange and lemon trees in the Bermudas were attacked by a minute insect, a species of *Coccus*, which caused sad havoc in the orchards. In a recent letter from a friend resident upon the islands, he gives the following particulars: "A few of the young lemon trees have escaped the disease, and some of the orange trees have also been saved by the simple means of cutting the tree down within two feet of the ground, and washing the stem with soft soap and water. I observed this remedy, and performed the work myself, with four orange trees, and they are now shooting up finely; and we have reason to think that the destructive disease is fading away; but the cause or origin is perfectly unknown to us, except that it consists in a mass of the minutest insects, all of which appear to form a glutinous substance when the hand is rubbed over it. The island of Antigua, which, in former years, produced the best description of oranges, was visited by this blight, and when I was there, twenty-eight years ago, not an orange tree was to be seen. In these islands of Bermuda, the visitation seems to have passed or skipped over one or two parishes in its progress from east to west, which is somewhat a mystery."

Orange and lemon walking sticks are in great requisition among the officers stationed on the Islands, and fine specimens are eagerly sought after, to carry home as presents to friends.

Shaddock, (*C. decumana*). Cultivated in gardens.

Locust tree of the West Indies (*Hymenæa coubaril*).

Tamarind (*Tamarindus Indica*). Grows to a large size. Very fine specimens at Walsingham.

Pride of India tree (*Melia Azedarach*). Generally planted near houses for ornament.

Calabash tree (*Cresentia cujete*). Moore's pretty lines on his favourite calabash tree at Walsingham, have caused that specimen, which is an extremely fine one, to be considered one of the "lions" of the Bermudas.

Bermuda Cedar (*Juniperus Bermudiana*). Male and female. Although considered a distinct species, it is, nevertheless, identical with the Virginian Cedar.

The beauty of the 'Mudian scenery is greatly enhanced by the cedar trees; as without them the Bermudas would present but a desert appearance, for it appears to be the only tree that can withstand the fury of the elements which, at particular seasons of the year, wage war upon those Islands. It affords great protection to the agriculturist, and stems the fury of the gale from his crops; it is of such great assistance in the proper tillage of the land, that the local Government have, at various times, passed Acts for the better protection of the tree from destruction. Much of the household furniture is manufactured from this wood, and room doors look extremely handsome when made of the knotted portion of the old cedars, and varnished. The beams of the houses, window frames, indeed, we may say that nearly all the wood required for building or house purposes, is cedar. Then, again, the far-famed 'Mudian boats are built of this wood, and not boats alone, but vessels of large tonnage are now and then, at intervals, launched in the Islands. Some small vessels of war have, at different times, been built also; one, named "The Bermuda," was launched in Harrington Sound some few years back, but she was the last man-of-war built of this wood, as it was found not to answer well for vessels of that class. In sheltered situations the cedar attains a goodly size, and there are yet remaining a few specimens of large dimensions, and supposed great age. There is one situate in the old churchyard of Smith's parish, which, from appearance, numbers many winters; it is put to an useful and decidedly ornamental purpose, for up in its aged and time-worn branches, is placed the church bell. Devonshire church owns a similar structure. The wood of the cedar is also used for fuel, and, to a stranger, the peculiar and delightful perfume emitted while the wood is burning is novel and pleasing. Perhaps we may here be permitted to venture an opinion, that the immense number of cedar trees add greatly to the heat of the climate, which, in the months of July, August and September, is generally intense. To judge of the heat thrown out by one of these trees, a person need only stand for a few minutes under its shade, when its influence will immediately be perceived. Many of the cedars growing close upon the shore, and exposed to the influence of the heavy gales, have a stunted appearance, the trunks and branches twisted into curious shapes, and their roots bare and exposed. In olden time the native "squires" of Bermuda calculated their wealth by the number and growth of the cedars upon their estates; and tradition has it, that the fair 'Mudian damsel who possessed a right and title to a thousand goodly cedars, was in a certain way of possessing a husband, should she feel so inclined.

Cocoa nut (*Cocos nucifera*).

Date palm (*Phœnix dactylifera*).

Cabbage palm (*Areca oleracea*).

Palmetto (*Chamærops excelsa ?*) This is the tall growing species. Grows well in the marsh near Brackish Pond Church.—[*See woodcut at the commencement of this paper.*]

Dwarf palmetto of the West Indies (*C. glabra*). Grows very freely in most situations. It is from the leaf of this tree that the celebrated "Bermuda plait" is made.

Mangrove (*Rhizophora Mangle*). Common near the shores, and in marshy tracts containing brackish water.

Avocada Pear (*Persea gratissima*).

Weeping willow (*Salix Babylonica*).

Butter nut (*Juglans* ——— ?) Called by the inhabitants the "Otaheite walnut."

Mulberry (*Morus alba*). Planted for silkworms. There is another species planted in the low part of the Court-house inclosure at Hamilton.

Tamarisk (*Tamarix gallica*), called in the Bermudas, "Spruce."

Pomegranite (*Punica granatum*). The double-flowering variety is also common.

Common purple Fig (*Ficus virens*).

Small white early Fig (*F. nymphæifolia ?*).

Peach (*Amygdalus* ——— ?), said to be an importation from Madeira. A hard fleshy variety is known as the "cling stone peach."

Loquat Tree (*Amygdalus* ——— ?). Recently introduced from China. It thrives and bears fruit admirably.

Mango (*Mangifera Indica*). In Mr. Ballinghall's garden, where the fruit ripened in 1851. Trees imported in the

French frigate, "Hermione," wrecked on the Bermuda Reefs. It also grows well, and produces fruit in Provost Marshall Trott's grounds, at Verdmont.

Round-leaved Sea-side Grape (*Coccoloba unifera*).

Papaw (*Carica papaya*). Male and female.

Oleander (*Nerium Oleander*). The double-flowering variety, very common, and very handsome. It is known to the natives by the name of "South-sea Rose."

Palma Christi, Castor-oil plant (*Ricinus communis*). Wild.

Fiddle-wood tree (*Cytharexylum cinereum*).

Banana (*Musa sapientum*). There is also a dwarf variety. The common banana is frequently stewed for dessert, and is delicious eating in that state.

Plantain (*M. paradisiaca*).

Prickly Lantana, or large red Sage Bush (*Lantana aculeata*). Common in garden hedges.

Barbadoes Flower Fence (*Cæsalpinia pulcherrima*). In gardens.

Red, or sweet-scented Jasmine (*Plumeria rubra*). In gardens.

Poison Weed (*Euphorbia laurocerasifolius ?*). A handsome creeping plant, not unlike the American creeper in appearance, climbing trees to the height of thirty feet or more.

Changeable Rose (*Hibiscus mutabilis*).

China Rose Hibiscus (*H. rosa Sinensis*). The "coral flower" of Bermuda.

Syrian Hibiscus (*H. Syrianus*). In gardens. Sometimes called "Althœa."

Common Ivy (*Hedera helix*). Recently introduced. Grows freely on northern aspects, and among trees.

Prickly Pear (*Cactus opuntia*). Grows wild all over the islands.

Cochineal Plant, or Indian Fig (*C. cochinillifer*). In gardens.

Scarlet-flowered Sage (*Salvia coccinea*). Road sides.

Common Sage Bush (*Lantana salvifolia*). All over the islands, wherever the land remains uncultivated. It forms the natural underwood of the islands.

Bamboo Cane (*Arundo bambos*). In lofty clumps.

Cultivated Reed (*A. fax*).

Arrow-Root (*Maranta arundinacea*). Few there are in England, who have not heard of the far-famed Bermudian Arrow-root, so highly prized as a light digestible food for invalids. The following remarks upon its cultivation and manufacture, by a gentleman resident in the islands, are too valuable to be omitted:—"This plant, as tradition informs us, was brought here from Charlestown, South Carolina; and, from its name, 'Indian Arrow-root,' must, I think, have been called so, from the resemblance of the end of the root, to the arrow used by the native tribes. The root, or stick, grows in the following shape, and the stem varies from two to four feet in height, the root seldom or ever being over an inch and a half in diameter, at the biggest, *or lower* end. I have seen as many as eighteen, and twenty sticks, or roots, growing from one stem. There is much trouble in digging this plant up, which process takes place during the months of January, February, and March. The root, at each joint, is covered with a skin, which is of a thick substance, and, as the root ripens, resembles a bit of gauze, and, when quite ripe, comes off almost as soon as

exposed to the sun; consequently, the buyer is benefited by purchasing the stick at a later season, say February and March, for then he buys, as you will readily see, *root*, instead of shoot or skin; but, as the means of manufacturing it into starch is somewhat limited (there being only about three machines of any power on the islands), the manufacturer has to begin early, and takes his chance of getting it with or without the skin or covering.

"The root is grated by a machine, worked by horse power; it is afterwards washed, and passed through sieves, or cloths, 'small by degrees, and beautifully less,' till all the flour is taken from it. It formerly took some time to dry, and would, occasionally, get mouldy; but, latterly, by the assistance of the screw press, it is instantly freed from water, and a day's drying in the sun renders the starch fit for use. There is less of it grown in Bermuda than ever, and, notwithstanding the price is much higher, and higher still it doubtless will be, it is not near so profitable as other things. It lies a whole year in the ground; potatoes only about a hundred, or a hundred and ten days, with much less trouble attendant on their culture, and a certain market in the United States."

That the Bermudian Arrow-root will, ere long, cease to be an important export from the Colony, is not to be doubted; and, taking into consideration the present condition and high price of labour, other products will be found a much more profitable speculation, and so cause "*real* Bermudian Arrow-root" to rank in scarceness with "*real* old East India Madeira."

In 1851, the value of the Arrow-root exported from the Bermudas, was £5,595 18s. 6d.

There is also a giant species of Arrow-root (*Canna coccinea*), the product of which is known by the name of "Tous le Mois."

Curascoa Swallow-wort, or Bastard Ipecechuana (*Asclepias curassavica*). The favourite food of the caterpillar of *Danais archippus.*

Mexican Argemone, or Prickly Yellow Poppy (*Argemone Mexicana*). The native Bermudians dye silk ribbons, &c., with the blossoms of this plant.

Iris-leaved Sisyrinchium, or Bermudiana (*Sisyrinchium Bermudiana*). Wild among rocks and sage bushes.

Azure Convolvulus (*Convolvulus nil*). A brilliant annual, spreading over trees of eight, or ten feet in height. Seeds abundantly.

There is a long trailing species common on the sea shore, with a blossom somewhat similar to *C. major*, known by the name of "bay bean."

West India Vervain (*Verbena Stachytarpheta*).

Grape Vine (*Vitis vinifera*). White and red.

Great American Aloe (*Agave Americana*). Blossoms freely.

Barbadoes Aloe (*Aloe Barbadensis*). The bitter juice of this plant is sometimes prepared for exportation.

Common Sugar Cane (*Saccharum officinarium*). In gardens only.

Yucca glorioso, "Spanish Bayonet," with its splendid panicles of white bell-shaped flowers.

The large Marsh Fern—growing among the mangroves of Chief Justice Butterfield's pond.

The common Fern—to be met with everywhere.

The Maiden's-hair Fern—which adorns the shady side of

road cuttings, where moisture filters through the rocky strata.

Lauristinus (*Virburnum tinus*).

Virginia Tobacco (*Nicotiana Tabacum ?*). In the wild state, by the sides of roads.

Red Lily, of Bermuda (*Amaryllis equestris*).

Guernsey Lily (*A. Sarniensis*).

Atamasco Lily (*A. Atamasco*). White blossom; fragrant.

Tournefortia, or Night-blowing Jasmine (*Nyctanthis* ——— *?*).

Crab Grass (*Agrostis Virginica*). A trailing species.

Apple, of Europe. Occasionally bears fruit. The tree, however, is very little known in the islands.

Many other trees and plants, natives of the West Indies, and elsewhere, might be enumerated, such as the Olive, Surinam Cherry, Scarlet-blossomed Cordia, Queen of Shrubs, Passion Flower, Myrtles, Yellow and White Jasmine, Roses, of many kinds, Cacti, Lilies, Custard Apples, Shell Plants, Heliotrope, Verbena, and a variety of garden flowers.

The Fuschia is frequently imported from New York; but the climate appears to be unfavourable to this plant; for although it will blossom in the shade of a northern aspect, it is never known to thrive—a singular fact, considering that hedges are formed of it in Madeira; the latitude of the two islands being the same.

The Lavender and English Wallflower both grow very freely, but refuse to blossom; and yet the German Double Stock, when not destroyed by the aphis, blooms magnificently.

The English Gooseberry, Currant, and Raspberry, run

entirely to wood, producing no fruit. The Strawberry bears sparingly. Melons thrive well, when the season is not too dry. The common sorts are very inferior. Water Melons, both round and oval, grow well. Cucumbers produce well, if sown early. The Bottle Gourd, introduced from Malta, by the 42nd Highlanders, grows remarkably well.

Vegetables, of many kinds, viz.:—Onions, Potatoes,* (common and sweet), Pumpkins, Cabbages, Carrots, Turnips, Parsnips, Tomatoes, Lettuce, Radishes, Leeks, Challots, &c., do well, considering the climate, though the cabbage kinds are sadly perforated by grubs and caterpillars. Indian corn of several varieties is grown, chiefly for use as a vegetable, and Guinea corn for green fodder. Barley begins to ripen in April; and the small patches of this grain are sometimes harvested by women, who clip off the ears with scissors, and carry them in their aprons. Oats are seldom, if ever, known to produce seed. Wheat produces a large, indifferent, ill-filled, husky grain, and is not sown for crop.

Hemp, sown in gardens, comes to perfection, and seeds well, much to the satisfaction of the red birds. Flax grows wild by the road side.

The Yam (*Dioscorea alata*), is not cultivated in the Ber-

---

* Large quantities of onions and potatoes are annually exported from the Bermudas; the former, principally to the West Indies; the latter, to the United States. Good prices are usually obtained for these vegetables; and, doubtless, if labour were more abundant, and the landed proprietors could only be persuaded to till the best portions of the land, now covered with cedar and sage scrub, they would eventually become much richer men, for they have a ready market for both these easily cultivated roots.

We forwarded a package of seeds of various kinds of English vegetables, about a year ago, to a friend resident in the Islands, and he gives us a very favourable account of their growth and yield; so we presume that in spite of drought, hurricanes, &c., most vegetables and fruits, whether boreal or tropical, will do well in the Bermudas.

mudas. Rice has been grown, by way of experiment, in wet and swampy ground, with success.

Capsicums, and peppers, of various kinds, do well. The edible Cassava (*Janipha manihot*), is common in gardens.

A species of Sanfoin, sometimes grown in the Bermudas, is the Spanish variety (*Hedysarum onobrychis*), introduced from Malta; it is deeper coloured than the common British species. It blossoms in the beginning of April.

The vignette below represents the inscription carved on the smooth indurated crust of a low rocky cliff on the South-shore, near Peniston's ponds, better known as "Spanish Rock." Tradition asserts it, with what truth we know not, to be the work of one Ferdinand Camelo, who is said to have paid the island a visit at the date marked on the stone. We are indebted to Mr. Hurdis for the copy of the inscription.

* See page 178.

FLATT'S BRIDGE.

METEOROLOGICAL OBSERVATIONS, MISCELLANEOUS NOTES, ETC.

THE Bermudas have long been considered as having an equable temperature by those who have never had an opportunity of judging by personal experience, and this wrong estimation of the climate has been greatly increased by the information of early as well as modern travellers and writers, who, judging merely from the occurrences of a brief residence, and, without the aid of meteorological and other observations, have given to the world at large an erroneous estimate of the "perpetual Spring" of those Islands.

Although fully aware of, and anxious to do justice to, the pretty and fairy-like scenery of those remote islands, yet we feel it would be totally inconsistent with a due regard for truth, were we, while descanting upon their merits, to pass over without remark, the disadvantages of the Bermudian climate.

During the hot months of summer the ground absorbs heat so largely, as to cause, by radiation, throughout the night, a temperature within three or four degrees of that of noonday.

It is not alone to the low level of the islands, to the high temperature by day and night, or to the frequent absence of wind, that the climate at this season of the year is so depressing, but to the dripping nature of the south-west wind, which then prevails and saturates the atmosphere with moisture to such an extent, as to cause it to resemble a vapour bath. Persons who have resided many years in the West Indies, under the refreshing influence of the dry East or Trade wind of those latitudes, acknowledge that the summer climate of the Bermudas is far warmer to the feelings, and far more enervating in its character, than any thing they have experienced to the Southward.

Now, in order to show the real features of the climate of the Islands, and trusting, at the same time, thereby to render this little volume as complete as possible, we are induced to insert the following observations made at the Royal Observatory, at St. George's, under the charge and superintendence of the Royal Engineers.

We consider it will be quite sufficient for the purpose we have in view, if we merely insert the observations made during the months of January and February, and July and August of last year, 1857, as the two former may be truly considered to represent winter, and the two latter summer temperature.

But in order more fully to show the variable nature of the 'Mudian climate, we think it as well to insert also the thermometrical and miscellaneous observations for each month of the year 1854, taken by our friend Mr. Hurdis.

The Bermudas are subject, but fortunately at rare intervals, to shocks of earthquake, which hitherto have happily done no harm beyond frightening the inhabitants, and thus interrupting the usual routine of their peaceable existence. An account of the last visitation of this nature, which took place on the 2nd of March of the present year, will be found in the following pages.

Hurricanes occur at intervals during the latter part of summer, and heavy gales very often during the winter months, doing considerable injury to trees, crops, and habitations, the former being uprooted, or having their branches stript of foliage, and torn asunder by the fury of the storm. The great hurricane which occurred on the 11th of September, 1839, did considerable damage to property of all descriptions.

EXTRACT FROM REGISTER KEPT AT THE ROYAL ENGINEERS' METEOROLOGICAL OBSERVATORY,

At St. Georges, Bermuda, during the Months of Jan., Feb., July, Aug., 1857. LAT. 32° 22' 57" N.—LONG. 64° 40' W.

JANUARY. Time of Observation 9½ a.m. Height above the Sea 122·64 feet.

Day of Month.	Barometer corrected & reduced to a temperature of 32° Fahrenheit.	Dry Bulb Thermometer.	Wet Bulb Thermometer.	Temp. of Dew point computed.	Elastic Force of Vapour.	Humidity 0—10	WIND. Direction.	Self Registering Thermometers. During Previous 24 hours. Min. in Grass.	Max. in Air.	Min. in Air.	Max. wet bulb.	Min. wet bulb.	Rain in previous 24 hours.	REMARKS.
1	29·924	62·8	61·3	60·3	·528	·920	SW b W	52·5	71·5	61·5	68·3	60·	Nil	
2	29·890	61·7	55·5	51·2	·388	·702	N b W	55 5	69·	61·	67·2	55·5	·12	
3	29·951	57·3	54·3	52·2	·403	·843	S b E	49·	63·3	54·	58·3	51·	Nil	
4	29·606	59·	52·1	47·3	·340	·672	WNW	53·	66·5	58·	65·	52·5	·55	Fresh Gale all day
5	29·843	56·5	50·7	46·6	·333	·714	NW	48·	60·2	55·5	56·	49·	Nil	Squally and showery all day
6	29·768	55·9	51·9	49·1	·362	·793	NW	50·	59·8	55·5	56·6	50·7	·07	Ditto ditto
7	29·807	52·	48·5	45·	·315	·789	NW	46·	57·2	51·	55·	48·2	·25	Ditto ditto Fresh Gale
8	29·711	53·	48·5	44·	·304	·734	W	45·	54·	49·8	52·	47·5	·08	Ditto ditto Strong Gale
9	29·881	49·	45·5	41·7	·280	·778	NNW	41·	54·8	46·2	52·5	42·	·17	Ditto ditto all forenoon
10	30·022	55·9	51·5	48·4	·354	·775	W	45·5	56·4	49 5	63·5	46·5	·04	
11	——	——	——	——	——	——		51·8	64·	56·8	61·	51·5	Nil	Strong breeze and showery all day
12	29 838	57·7	54·5	52·8	·404	·834	NW	50·6	65·6	58·2	64·	54·	·23	
13	30·074	54·1	50·1	46·1	·327	·761	N b W	48·	58·8	52·5	56·7	47·5	Nil	
14	29·658	59·	57·3	56·1	·460	·908	ESE	52·5	59·4	54·4	58·	50·	·45	Fresh gale changing suddenly from E.S.E.
15	29·945	60·3	55·5	52·1	·402	·760	W b S	47·5	62·7	56·2	62·	54·7	·11	[to N. N. W. at 12½ p. m.
16	30·071	52·8	47·8	42·8	·291	·709	N	45·	65·7	53·2	61·	48·	·04	Strong breeze all day
17	30·239	53·2	49·5	45·8	·324	·777	SW	46·8	54·5	51·2	51·3	45·5	Nil	Nearly a calm till noon
18	30·035	61·	56·3	53·	·414	·765	SSW	56·8	61·5	58·8	57·5	50·	Nil	
19	29·663	65·5	63·5	62·3	·565	·898	S	55·	66·	61·	64·	56·	·05	High wind, thick fog and rain nearly all
20	29·796	51·1	48·3	45·5	·320	·825	WNW	47·	66·	50·8	64·5	48·6	·35	Squally and showery all day [day.
21	30·255	53·5	49 5	45·5	·320	·760	WSW	47·8	63·3	50·8	56·5	46·5	·02	Nearly a calm all day [tween 11 a.m. & 1 p.m.
22	29·784	62·9	58·	54·6	·436	·757	SSW	54·8	62·8	53·8	58·	49·5	Nil	High wind at 9½ a.m. rising to brisk gale be-
23	29·953	54·8	51·5	48·2	·351	·799	WNW	51·	64·5	55·	63·	51·3	·08	Fresh beeze and passing showers
24	30·296	51·5	49·	46·5	·331	·843	NW	46·	58·	50·6	54·	47·	·02	Beautiful weather
25	30·221	59·7	54·2	59·4	·378	·730	NNW	53·	61·5	51·3	58 3	48·7	·08	Ditto
26	30·227	58·5	55·7	53·7	·424	·852	N b W	52·	65·	57·5	60·	53·5	·07	
27	30·321	55·5	50·5	47·	·337	·749	NE	47·7	60·	52·	58·3	46·	·02	
28	30·008	60·9	58·7	57·1	·476	·883	NE	52·5	61·8	55·	59·5	50·	Nil	Fresh breeze all day
29	29·726	62·1	61·9	61·8	·555	·989	NNE	58·5	63·5	59·6	62	59·	·14	Showery all forenoon
30	29·928	63·	59·5	57·1	·475	·820	NNW	59·3	65·5	61·1	68·8	58·9	·01	Very pleasant weather
31	30·156	60·3	58·3	56·9	·471	·893	NNW	50·2	66·5	56·8	62·5	55·5	Nil	

Barometer—Highest 30·321 at 9½ a.m. on 27th. Lowest 29.586 at 3½ p.m. on 4th. Total fall of Rain 2·93 inches.

The following Means are deduced from an average of four daily observations, taken at 3½ and 9½ a.m. and 3½ and 9½ p.m. :—

Mean Barometer	29·929	Mean Temp. of Dew Point	51·4
Mean Thermometer (Dry Bulb)	57·8	" Elastic force of Vapour	3·91
" ditto (Wet Bulb)	54·2	" Humidity	·808

FEBRUARY. (4h. 18' 40" Time.)

Day of Month.	Barometer corrected & reduced to a temperature of 32° Fahrenheit.	Dry Bulb Thermometer.	Wet Bulb Thermometer.	Temp. of Dew point computed	Elastic force of Vapour.	Humidity 0—1	WIND. Direction.	Self registering Therms. Min. on Grass. (During previous 24 hrs.)	Max. in Air.	Min. in Air.	Rain in previous 24 hours.	REMARKS.
1	30·104	63·5	60·3	57.7	·488	·831	S b E	57·5	68·	59·7		Raining until Noon. Pleasant afternoon
2	29·743	62·3	62·0	61.7	·546	·946	SSW	60·2	70·	62·4	.2	Pleasant weather. Light rain at 3½ a.m.
3	30·092	59·6	55·5	52.2	·404	·774	NW b N	54·8	66·3	58·5		
4	30·317	56·0	50·5	45.6	·333	·726	NNE	50·	65·5	53·		
5	30·396	58·1	53·2	48.79	·371	·760	NE	52·8	59·8	55·8		
6	30·261	63·3	61·	59.4	·507	·858	ENE	52·	64·	55·7		
7	30·202	64·9	63·3	62.4	·569	·918	E	60·	65·	63·4		
8	30·231	65·8	64·5	63.5	·588	·933	ESE	58·	57·5	64·		Pleasant weather
9	30·213	65·5	64·5	63.4	·594	·947	SE	56·5	70·6	64·2		Close dull day
10	30·114	66·3	65·7	65.0	·622	·958	SE	60·5	72·	64·8		Raining at 3½ a.m.
11	30·129	58·3	53·5	49.7	·375	·759	N b E	53·	70·8	57·3	·34	Ditto
12	30·396	53·3	51·3	49.3	·367	·877	NE	47·	71·	52·		Squally and showery all day
13	30·407	60·2	58·2	56.8	·470	·893	E	49·5	61·	53·3		
14	30·231	62·3	61·3	50.2	·534	·946	E b N	47·	64·	60·	·03	
15	30·239	62·5	58·0	54.8	·445	·776	ENE	57·	67·	59·2		
16	30·207	63·5	60·7	59.5	·496	·824	E b N	56·	67·2	61·2		Light showers during the day
17	30·129	64·4	63·9	63.5	·591	·973	E b N	54·2	70·	62·9	·075	Light rain all day
18	29·922	63·5	60·7	59.6	·516	·893	ENE	57·5	67·5	62·8	·02	Thick fog and drizzly nearly all day
19	29·850	64·5	64·2	64.0	·522	·973	NE b E	58·5	66·2	61·8	·07	Drizzle rain at 9½ p.m.
20	29·912	63·9	62·5	61.5	·550	·924	NE	57·	68·8	61·5	·02	Ditto
21	30·022	65·3	63·9	63.0	·579	·927	NE	53·2	68·5	61·5		Ditto 3½ a.m.
22	30·020	64·1	61·1	60.0	·507	·845	WNW	53·6	71·	61·4		Hazy all day
23	29·976	65·6	64·2	63.4	·585	·927	W	59·	70·6	64·1		
24	30·153	63·7	56·5	51.6	·393	·668	N	56·2	69·8	60·7		Pleasant weather
25	30·256	62·9	57·5	53.8	·424	·765	S b W	50·7	67·5	58·		Raining in the afternoon
26	29·871	66·9	62·9	60.2	·532	·810	SW	58·	67·8	62·3		Pleasant weather. Fresh breeze at 3½ a.m.
27	30·072	56·8	50·5	46.2	·328	·700	N b W	53·	63·4	56·	·05	
28	29·951	60·2	55·7	52.5	·407	·774	SW b S	50·	60·5	55·		Lightning to the North-westward.

Maximum Barometer during the Month of February, on the 4th at 9½ p.m. 30·411. Total fall of rain ·805.
Minimum " " " 2nd at 3½ a.m. 29·739.
The following means are deduced from an average of four daily observations at 3½ and 9½ a.m. and p.m.;—

	A.M. 9·30	A.M. 3·30	P.M. 3·30	P.M. 9·30	General Means.
Mean Barometer	30·121	30·107	30·099	30·118	30·111
Mean Thermometer Dry Bulb	62·4	64·4	57·86	61·6	61·5
Mean Thermometer Wet Bulb	59·6	60·9	58·02	59·2	59·205
Mean Temperature of Dew point	57·09	58·5	56·15	57·4	57·28
Elastic Force of Vapour	·586	·509	·472	·494	·490
Humidity	·846	·880	·927	·878	·870

JULY. Time of Observation 9·30 a.m.

Day of Month.	Barometer corrected & reduced to a temperature of 32° Fahrenheit.	Dry Bulb Thermometer.	Wet Bulb Thermometer.	Dew point computed.	Elastic force of Vapour.	Humidity 0—1	WIND. Direction.	Self Registering Thermometers. During previous 24 hours. Max. in Sun.	Min. on Grass.	Max. in Air.	Min. in Air.
1	29·652	78·2	72·7	70·0	·725	·767	w b s	85·8	66·4	83·2	74·2
2	29·885	78·2	77·5	72·7	·791	·837	ssw	84·1	70·0	81·	74·
3	29·979	78·8	76·2	74·7	·857	·932	s	80·8	65·	79·5	72·
4	30·075	79·5	76·9	74·5	·849	·883	s b w	83·8	74·	80·4	75·
5	30·178	80·2	77·5	76·2	·877	·880	SSE	87·7	74·	82·	76·5
6	30·247	78·9	75·9	73·6	·818	·884	E b s	94·5	71·5	84·	74·
7	30·158	78·0	73·6	71·4	·761	·809	SE	98·8	70·2	83·	73·
8	30·059	79·0	75·	73·0	·801	·825	SSE	96·8	70·	83·	73·
9	30·107	81·5	77·3	75·2	·860	·833	E b s	101·6	71·7	84·1	75·6
10	30·179	81·3	76·7	74·4	·838	·806	NE	102·5	70·	84·8	75·
11	30·169	80·0	76·	74·0	·827	·826	SSE	104·8	72·	86·	76·
12	30·004	79·2	75·	73·1	·798	·817	wsw	96·1	72·	83·5	76·
13	29·914	76·9	73·9	72·2	·789	·867	NW	91·5	78·	81·8	75·
14	29·926	78·	75·3	74·0	·826	·878	s	89·6	79·	80·5	72·8
15	30·017	76·1	72·5	70·7	·746	·841	s b E	101·5	71·	81·2	73·
16	29·984	77·3	75·3	74·3	·835	·908	SE b E	82·9	69·2	78·3	72·6
17	29·973	78·8	77·3	76·6	·898	·977	s	93·7	72·	80·8	75·4
18	29·962	79·1	76·9	75·8	·877	·901	sw	85·7	66·	82·	76·8
19	29·933	80·5	77·3	75·7	·874	·875	sw	89·2	73·	84·	76·5
20	29·984	80·8	77·5	75·9	·878	·888	sw	87·7	76·	83·5	77·3
21	30·046	80·2	76·8	75·1	·817	·852	ssw	87·	77·	81·5	76·2
22	30·102	79·3	76·3	74·8	·849	·867	s	87·5	72·4	83·	76·2
23	30·107	80·5	77·3	75·7	·874	·877	SSE	87·9	73·	83·	77·2
24	30·187	81·5	77·5	75·5	·868	·826	ESE	95·	72·8	84·	76·6
25	30·207	81·2	76·4	74·0	·827	·794	ENE	104·5	72·7	87·2	76·6
26	30·173	80·1	74·5	71·7	·769	·765	E b N	99·5	70·	85·4	75·
27	30·117	81·9	76·9	74·4	·837	·787	NNE	100·	72·	84·8	76·
28	30·148	81·3	77·	74·8	·850	·814	NE	104·8	71·6	81·5	76·
29	30·176	82·2	77·5	75·2	·858	·799	s	97·5	73·8	88·	78·
30	30·167	82·	77·3	74·5	·868	·819	SSE	94·9	74·5	89·5	78·
31	30·138	81·5	76·5	74·0	·827	·788	s	95·5	74·1	88·	78·

Maximum Barometer during the month of July on the 6th at 9·30 a.m., 30·247.
Minimum " " on the 1st at 3·30 p.m., 29·652.
The following Means are deduced from an average of four daily observations taken at 9·30 and 3·30 a.m., and 9·30 and 3·30 p.m.

Mean Barometer		30·087
Mean Thermometers	Dry Bulb	78·6
	Wet Bulb	75·2
Mean Temperature of Dew point		78·4
" Elastic Force of Vapour		·797
" Humidity		·825

AUGUST. (Time 4h. 18m. 40s.)

Date.	Barometer corrected to a temperature of 32° Fahrenheit.	Thermometer. Dry.	Thermometer. Wet.	Dew point computed.	Elastic force of Vapour.	Humidity.	Self Registering Therms. Max. and Min. during previous 24 hours. Max. in Air.	Min. in Air.	Min. on Grass.	Max. in Sun's Rays.	Wind.	Rain in previous 24 hours. On Ground.
1	30·011	81·7	76·6	73·7	·809	·766	88·	77·8	82·	92·8	s b w	—
2	30·128	80·5	75·	72·3	·781	·768	86·	77·5	70·	91·	s w	—
3	30·025	82·5	77.2	74·5	·842	·777	86·	78·	77·8	92·2	w b s	—
4	30·033	80·5	75.5	73·0	·800	·786	85·8	78·2	71·6	90·8	w	·05
5	30·058	79·3	75.5	73·6	·817	·834	86·	74·	66·6	88·5	w b N	0·26
6	30·098	81·5	76.5	74·0	·827	·788	86·5	77·5	72·	98·5	s b w	—
7	30·185	81·5	77.5	75·5	·868	·826	87·	77·	71·8	93·6	SSE	·03
8	30·186	83·	77.3	74·5	·839	·762	88·	76·3	69·3	106·6	E	—
9	30·073	82·2	76.9	74·3	·833	·776	88·	76·3	71·	106·4	SE	—
10	29·962	79·8	75.3	73·5	·802	·806	88·4	75·	70·1	93·3	WSW	·07
11	29·988	81·3	75.5	72·6	·791	·757	86·	78·	80·	91·8	SW b s	—
12	30·060	79·2	76.9	75·7	·875	·897	85·	77·5	82·	92·2	SW b s	·02
13	30·097	82·3	77.1	75·0	·840	·780	84·8	77·6	74·	90·8	NW	—
14	30·633	80·5	75.1	72·4	·785	·772	86·3	76·3	72·4	98·5	N b w	—
15	29·945	82·1	76.7	74·0	·827	·772	86·	78·	74·	97·7	NW	—
16	29·992	81·5	76.5	74·0	·827	·788	88·	78·2	72·6	96·95	W	—
17	30·011	81·5	77.5	75·5	·868	·826	86·5	79·	75·	97·7	w b s	—
18	29·943	79·	77·3	76·4	·896	·922	87·4	77·2	74·	91·7	w b N	·30
19	29·898	82·	78·	76·0	·882	·827	83·8	79·2	76·	90·	NW	—
20	29·930	82·3	78·3	76·3	·891	·827	87·	78·	73·8	94·8	S	—
21	29·965	80·7	78·5	77·4	·922	·901	85·	79·	75·7	91·2	SSW	·17
22	29·965	77·9	76·3	75·5	·840	·842	82·8	76·8	72·2	92·7	WSW	·39
23	29·960	82·8	78·1	75·8	·875	·799	85·	78·5	74·6	105·8	SSW	—
24	30·036	82·3	77·3	74·8	·849	·788	87·3	79·	76·	91·5	SW b s	—
25	30·098	76·1	74·9	74·3	·835	·964	85·	76·	73·	90·5	W	·33
26	30·096	79·	74·3	71·9	·774	·798	81·	74·2	71·	86·	NNE	1·58
27	30·003	77·2	70·5	67·2	·663	·723	83·4	74·6	72·	86·90	NE	—
28	29·944	79·	71·1	67·1	·662	·681	82·	73·8	68·	95·5	NNE	—
29	29·866	79·5	74·9	72·6	·790	·802	84·	74·9	70·	100·8	E b N	—
30	29·906	79·7	75·1	73·5	·795	·802	84·8	76·8	71·5	101·4	SSW	—
31	29·934	80·1	76·5	74·7	·846	·843	82·	75·	70·	87·	NNW	·96

The following Means, Maxima and Minima are deduced from four daily observations taken at 9·30 and 3·30 a.m. and p.m.

Mean Barometer	29·996	Max. Bar. 30·186 at 9·30 a.m. on the 8th
Do. Thermometer (Dry)	79·9	Min. do. 29·853 at 3·30 a.m. on the 30th.
Do. do (Wet)	75·6	Max. Ther. 87·7 at 3·30 p.m. on the 8th.
Do. Dew Point	73·5	Min. do. 73·5 at 3·30 a.m. on the 28th.
Do. Elastic force of Vapour	·812	
Do. Humidity	81·2	

DAILY JOURNAL OF TEMPERATURE.

REMARKS UPON THE STATE OF THE WEATHER—DIRECTION OF WIND—MISCELLANEOUS NOTES, &C., FOR THE YEAR 1854.

BY JOHN L. HURDIS, ESQ.

JAN.	8 a.m.	Mid-day.	6 p.m.	10 p.m.	REMARKS.	WIND.
1	67	72	67	67	Cloudy. 3 p.m. drizzling showers. Peach trees beginning to blossom. Ship in distress signalled	S.
2	68	72	68	60	Cloudy. 3 p.m. showery. a.m. wind west. 3 p.m.	NW.
3	60	57	56	56	Cloudy, with cold drizzling showers. 10 p.m. rain	NE.
4	60	62	65	63	A wet day. General thanksgiving in all the parishes on account of the disappearance of yellow fever	NE.
5	58	63	61	57	A bright pleasant day. The band of the 56th Regt., which lost eleven men during the fever, and is now only twelve in number, re-appeared in public, & played on the green near Dean's Academy	S.
6	58	66	60	58	9 a.m. raining. Remainder of the day fair. Strawberry plants beginning to blossom	W.
7	62	60	62	62	Raining for some hours. Wind W., then N. and E.	Variable
8	62	77	71	66	Heavy showers last night. Cloudy with strong wind. Lauristinus in blossom	SSW.
9	60	57	—	57	Lightning, thunder, and rain last night. To-day gleams of sunshine, with strong wind	NW.
10	51	59	57	54	Fair. English mail arrived per "Merlin." Band played	NW.
11	52	66	61	63	Fair. Missing English mail signalled. Orange crops gathered	E.
12	68	66	65	65	Fair. Coughs, colds, and sore throats very prevalent	SE.
13	68	70	67	66	Cloudy and damp. A case of small pox in Hamilton Harbour	SSW.
14	67	69	65	62	Dark and cloudy. 2 p.m. rain	SW. to W
15	60	64	54	51	Fair. Lemon trees loaded with ripe fruit, but almost denuded of leaves by the wind	N.
16	50	64	53	49	Beautiful day	N.
17	55	67	61	64	Fair. Sale of effects, the property of officers who died of yellow fever	W.
18	67	69	65	65	Fair	W.
19	65	68	61	61	Fair	NE.
20	61	67	64	64	Fair. Cultivators setting out onion plants for crop	SW.
21	65	68	64	64	Fair. Sundry gates and sign boards missing in Hamilton	SW.
22	66	70	55	55	Cloudy. at noon rain with strong breeze from N. between noon and 1·30 p.m. therm. fell ten degrees	SW. to N
23	61	70	67	59	8 a.m. heavy shower. then cloudy, with strong breeze. 8 p.m. more rain with distant lightning, the wind veering to N...	SW. to N
24	56	59	54	53	Fair. White rose trees full of buds	N.
25	55	60	57	57	Fair	N.
26	54	64	59	58	Fair. Coughs and sore throats still prevalent	E.
27	62	69	64	64	Showery until 9 a.m., then fair. At night strong breeze	SW.
28	66	63	60	57	Showers, with strong breeze till noon. p.m. cloudy	SW to NW
29	55	58	53	50	Cloudy and chilly. Heavy dew at night	NE.
30	51	51	54	58	A dark and dismal day; raining incessantly, which continued for the greater part of the night	NE. to S.
31	62	69	63	62	Fair ...	W.

NOTE.—This was a very stormy month, varied, as usual, by a few bright sunny days between each gale. *Hamilton* Harbour unusually full *of* vessels in distress. H.M. Schooner "Bermuda", from Saint Domingo, said to have sent a case of yellow fever to the Royal Naval Hospital at Ireland Island. The Crop of "sweet" oranges never before known to be so plentiful. Seed potatoes from the United States largely planted. Onion plants said to be a partial failure from the bad quality of the seed imported from Madeira. On warm sunny days the red bird, blue bird, and white-eyed greenlet, all sing a cheerful ditty in the cedars—and the former of these birds invariably at the very earliest dawn of day.

FEB.	8 a.m.	Mid-day.	7 p.m.	10 p.m.	REMARKS.	WIND.
1	60	65	56	53	Fair, a.m. wind N. p.m. S. 6 p.m. W.	Variable
2	53	65	57	58	Bright sunny day. Narcissus in bloom	W.
3	58	70	66	66	Cloudy, threatening rain	S.
4	57	56	50	50	Morning fair. Evening cloudy with showers. Blowing very freshly	NW.
5	50	55	50	45	a.m. showers p.m. fair. Cold bracing weather	NW.
6	52	57	55	55	Blowing a strong breeze, 3 p.m. rain, with distant thunder and lightning (2 p.m. therm. 60)	WSW.
7	46	55	52	51	A wintry day. Therm. indoors 60, with a bright fire burning *	N.
8	50	57	56	58	Cloudy and cold (English mail signalled), 10 p.m. wind S.	E.
9	62	69	—	63	Blowing half a gale at sunset, steady rain for some hours	S.
10	64	62	58	57	A thorough wet day. Raining all last night	SW.
11	60	63	56	52	Cold and windy	NW.
12	51	56	52	50	Cloudy, cold and windy. 10 p.m. light showers	N.
13	54	59	56	53	Fair, with cold breeze. Governor Elliot sailed for Barbadoes	N.
14	43	64	53	54	Fair and calm. 10 p.m. wind SW.	N.
15	48	65	61	60	Fair	W.
16	54	68	60	58	A summerlike day	W.
17	61	73	65	62	Fair. Gladiolus ——— ? in blossom	WSW.
18	57	55	55	55	Raining nearly the whole day	NE.
19	56	60	50	47	Fair	NW.
20	51	55	—	59	Cloudy, p.m. continued rain	SE.
21	66	70	65	62	Rain and wind. A strong S. wind and rain all last night	SW.
22	56	65	56	52	Fair. 10 a.m. showery	N.
23	51	65	60	57	Showery, with a strong breeze blowing	SW.
24	54	65	53	47	Fair, but chilly and cold out of the sunshine	NW.
25	60	61	52	48	Fair, calm	N.
26	57	65	—	61	Cloudy	SE. to S.
27	64	68	66	66	Cloudy, with strong breeze (11 p.m. rain)	SW.
28	60	64	58	58	Fair	NW.

* This degree of temperature must be distressing to the native population, who seldom indulge in fires, and who mostly sleep under a single blanket.

February is noted for boisterous winds and rainy weather, and a temperature, at times, so low, as to be almost unbearable without the aid of a good fire. Most of the dwellings in Bermuda are damp and cold from the absence of any means of warming them in winter. Coughs, colds, sore throats, fearful attacks of rheumatism, and even chilblains, are consequently prevalent at this season. Females envelop themselves in warm shawls throughout the day, and men take strong exercise to warm themselves. Potatoe planting is very general.

NOTE.—In the following year, 1855, the quantity of rain which fell during the month of February, as per Register kept at Ireland Island, by the late J. D. Anderson, Esq., C.E., was 9.34 inches.

MARCH	8 a.m.	Mid-day.	7 p.m.	10 p.m.	REMARKS.	WIND.
1	57	67	57	57	Fair and calm	NE to W
2	54	67	58	57	Fair	N
3	46	67	56	54	A beautiful bright sunny day	NW
4	60	70	62	62	Fair. A herring gull appeared in Hamilton Harbour	SW
5	65	65	61	61	11 a.m. rain. From 2 p.m. fair	W
6	59	65	59	59	Fair	SW
7	56	64	53	47	Fair	N.
8	55	67	63	62	Fair	S.
9	65	73	65	63	Fair	SW
10	66	75	64	63	Fair	SW
11	61	74	64	65	Fair, with strong breeze	SW
12	65	74	59	60	Cloudy. 4 p.m. rain	SW
13	61	65	62	60	Cloudy at intervals. 5 p.m. threatening rain	S
14	60	64	59	57	a.m. cloudy, and threatening rain. p.m. fair	SE
15	63	69	63	—	Showers before daylight. Fair rest of the day	S
16	65	70	67	67	Cloudy, with strong breeze	WSW
17	68	75	67	67	Cloudy, with thick mist, blowing freshly	WSW
18	68	73	68	68	Cloudy. Blowing hard. Boats close reefed	SW
19	57	62	58	57	Fair	NW
20	59	62	58	58	Fair, evening cloudy	WNW
21	53	57	59	55	Cloudy and cold. Slight showers at sunset	NW
22	50	52	51	49	Bright sunny day	NW
23	58	70	—	63	Cloudy. Raining more or less all day	S to SW
24	62	72	62	61	A beautiful day	W
25	62	70	64	65	Fair. Blowing strong breeze	SW
26	63	67	57	55	Cloudy. 7 p.m. rain, which continued for some hours, clouds setting from SW. and N. at same time	SW to N
27	56	60	55	53	Boisterous, with flying showers of rain	NW
28	57	65	—	60	Fair. Cold. At night showers	WSW
29	55	62	53	53	Cloudy and cold. Double German stocks in bloom and looking very handsome	NW
30	55	63	54	46	A beautiful day. Orange trees budded, sometimes on lemon stocks	NW
31	59	66	—	59	Fair. A comet observed last evening in the NW.	SE

At different periods during the month a fire became necessary to render the house habitable; at other times a summerlike temperature prevailed. But little rain fell during this month. Towards its close the native birds commence building their nests.

APRIL.	8 a.m.	Mid-day.	7 p.m.	10 p.m.	REMARKS.	WIND.
1	62	73	64	63	Fair. Evening cloudy	SE.
2	66	73	67	66	Fair with strong breeze. An osprey observed	S.
3	66	75	66	66	a.m. Fair. p.m. cloudy, with showers	SW.
4	60	64	60	58	Fair. Raining last night	N.
5	60	64	—	57	Cloudy. A kind of influenza prevailing in the neighbourhood	N.
6	60	—	55	49	Fair. Native birds in full song	N.
7	58	70	58	58	Fair. Strawberries ripe in the Admiral's garden	S.
8	61	—	58	53	a.m. Cloudy. At noon slight shower, p.m. fair	N.
9	58	67	60	60	Fair. Blue birds feeding their young	S.
10	66	70	67	66	Cloudy, with strong breeze. Slight showers at night	S.
11	68	73	68	64	Fair. Beautiful showers last night	SW.
12	65	65	60	57	a.m. Fair, p.m. cloudy	N.
13	61	69	60	57	A beautiful day	N.
14	58	68	58	54	Fair. Sweet peas in bloom	N.
15	65	73	65	65	Fair. Rain much wanted	S.
16	71	76	69	69	Fair, with strong breeze, and thick haze to windward	SW.
17	68	72	67	65	Dark and cloudy, with occasional showers. At night continued rain	S.
18	60	70	64	60	Cloudy, with light showers. At sunset fair. The red lily (amaryllis equestris) in bloom	S.
19	65	72	63	62	Splendid heavy showers 'till noon. p.m. fair	SW.
20	67	70	59	54	A beautiful clear calm day	NW.
21	62	72	65	65	Fair. White roses abundant	W.
22	66	70	56	53	Fair. Red lilly ditto	N.
23	61	78	67	64	Fair. Steamer "Curlew" from St. Thomas. Declaration of war with Russia	S.
24	58	68	58	56	Rain at an early hour, then fair, with strong breeze	NW.
25	60	65	60	59	Fair	W.
26	66	72	63	62	Fair. Multiflora rose in full bloom	W.
27	68	72	61	57	Fair. Mignionette blossoms	SW.
28	66	73	64	64	Fair. Patches of barley ripening	E.
29	66	72	63	62	Fair. Carting potatoes into town for shipment	E.
30	66	73	62	62	Fair. Sea bottles on the south shore	E.

MAY.	8 a.m.	Mid-day.	7 a.m.	10 p.m.	REMARKS.	WIND.
1	69	79	61	57	Fair	E.
2	66	75	64	64	Fair. Light airs	W.
3	66	72	62	60	Slight showers to 9 a.m. then fair	NW.
4	66	75	65	65	Fair. A slight shower at 8 a.m.	S.
5	65	—	64	64	Fair. Strong breeze	W.
6	69	75	65	64	Fair, ditto. Sweet peas very abundant	W.
7	68	75	64	64	Cloudy, threatening rain. At 7 p.m. thunder shower. Rain during the night	SW.
8	60	71	60	60	Cloudy, English mail signalled at 6 p.m.	N.
9	65	70	—	57	Beautiful day	N.
10	64	74	63	62	Fair	E.
11	66	73	66	64	Fair. Evening cloudy and threatening rain	E.
12	66	74	67	—	Cloudy, with thick haze and slight showers. Raining during the night	E.
13	69	74	68	67	a.m. Cloudy and wet, with strong breeze. p.m. cloudy	E.
14	69	74	68	65	Fair	E.
15	68	—	69	—	Fair. Red and blue birds with second nests	SE.
16	68	77	68	65	Fair. Brisk exportation of onions and potatoes to the United States & West Indies—the former chiefly to the West Indies	SE.
17	70	77	69	64	Fair. English mail arrived	S.
18	70	81	67	—	Warm summerlike weather	S.
19	70	81	70	64	Ditto ditto	S.
20	71	76	68	63	Fair	E.
21	68	—	68	64	9 a.m. Showers. Remainder of the day fair	E.
22	70	77	68	67	Fair	E.
23	69	79	68	67	Fair. General Assembly convened	E.
24	69	—	67	—	Fair. Regatta in Hamilton Water, and ball at the Mason's Hall	E.
25	70	79	65	58	Fair	N.
26	71	78	68	68	Fair. Ground very dry	WSW.
27	72	80	71	69	Fair, with strong breeze. 10 p.m. gentle showers and lightning from the north	SW.
28	73	77	66	65	Fair. p.m. flying clouds and light showers	NW.
29	71	76	70	68	Fair. White lilies in bloom	NW.
30	73	80	—	67	Fair. English mail arrived	NW.
31	74	—	72	70	Fair. A light shower at 11 a.m. Regatta in the Great Sound	SW.

JUNE.	8 a.m.	Mid-day.	7 a.m.	10 p.m.	REMARKS.	WIND.
1	75	80	73	72	Beautiful showers from 8 to 10 a.m., then fair, with a strong breeze	SW.
2	76	80	73	70	Fair	SW.
3	72	77	71	70	Fair. Sweet potatoe cuttings or slips planted	N.
4	73	79	76	73	Fair. p.m. cloudy	SE.
5	75	80	75	72	Fair	S.
6	77	81	73	73	Cloudy, 5 p.m. light showers	S.
7	75	81	75	73	Fair, Sky partially clouded. Ball at St. George's	S.
8	—	—	73	73	Cloudy, with strong breeze. 6 p.m. slight shower	S.
9	81	81	75	72	Fair. Cucumber plants withering from heat	S.
10	82	85	78	73	Fair. The shrill jarring note of the Cicada, or "scissar grinder" heard among the cedars	SW.
11	82	84	78	73	Fair. Second brood of young blue birds under the eaves of our cottage	SW.
12	83	84	73	75	Fair. Strong breeze	SW.
13	78	87	—	74	Fair. Evening cloudy	SW.
14	79	86	—	73	Splendid showers of rain at intervals. English mail arrived	SW.
15	73	—	77	—	Heavy rain with thunder & lightning before daylight. Tank overflowing. Sky considerably clouded	SW.
16	74	82	70	—	Fair. Honeysuckle in blossom	W.
17	75	82	78	76	Fair. Birds still nesting, Young red and blue birds abundant	NW.
18	75	80	—	70	Fair. Anniversary of Waterloo	NE.
19	75	80	75	74	Fair. Carried oat crop, sown 9th February and 7th March	E.
20	72	—	—	—	Heavy showers throughout the day	S.
21	81	80	78	72	Fair. The brown chafer or "hardback", very abundant	SW to NW
22	78	78	78	78	Fair. The green bug common	W.
23	78	80	—	—	Fair. Heavy shower in the night	W.
24	—	80	78	78	Fair. Showers at 4 p.m.	SW.
25	79	79	79	79	Fair. Shower 7 p.m. strong breeze	SW.
26	79	80	76	76	Fair. Heavy shower at 3 p.m. English mail arrived	W.
27	75	74	76	76	Dark and cloudy. 10 a.m. heavy rain with thunder and lightning. 6 p.m. pouring rain for an hour	SW.
28	76	80	75	75	Fair. Gathered first melons of the season	NW.
29	74	78	76	76	Fair	SW.
30	77	78	76	76	Fair	SW.

JULY.	8 a.m.	Mid-day.	7 p.m.	10 p.m	REMARKS.	WIND.
1	79	79	79	79	Fair. At noon a smart shower of rain	SW.
2	79	79	76	76	At 9 a.m. shower with distant thunder. Cloudy	N.
3	77	78	76	76	Fair. Sky partially clouded	E.
4	75	—	76	76	Fair. calm. Oleander hedges in full bloom	E.
5	75	—	76	76	Fair. Evening primrose in blossom	W.
6	78	84	80	79	Fair. Sweet scented tournafouchia in bloom	W.
7	79	83	79	—	Fair. The red bird still in song	NW.
8	80	84	79	79	Cloudy and sultry. At noon a thunder shower. Evening cloudy	WNW&NE
9	79	79	76	76	Fair. Red admiral butterfly has recently appeared	E.
10	76	78	78	78	Fair. English mail arrived	E.
11	80	83	79	79	Fair. Melons in full seasonm	SE.
12	81	84	78	78	Fair. Evening calm and sultry	S.
13	82	81	81	80	Fair, calm and sultry	Variable
14	81	83	80	80	Fair. Vegetation parching on shallow soils	N.
15	81	82	76	76	Fair. A few drops of rain fell during the day. Cloudy with distant thunder at sunset	E.
16	81	82	79	78	Fair. A blue wasp observed. Unfledged blue birds still met with	S.
17	81	84	80	79	Fair. At night lightning, distant thunder, and roaring of the sea, with showers of rain	S.
18	81	83	80	79	Fair. Strong breeze with occasional showers	SW.
19	81	84	80	78	Fair. Lightning and distant thunder at night	SW.
20	81	82	79	77	Fair, with strong breeze	SSW.
21	80	83	79	77	Fair. 3 p.m. heavy shower. Steamer from St. Thomas signalled	SW.
22	81	82	80	79	Fair	SE.
23	82	82	80	79	Fair, a light shower at 4 p.m.	SE.
24	81	82	80	79	Fair. Spotted sandpiper arrived from the north	SE.
25	82	82	81	80	Fair. The lesser Yellow shanks abundant at Peniston's ponds	SE.
26	81	83	80	76	Fair. At midnight calm. Therm. 75	SE.
27	81	82	80	74	Fair, warm and sultry. At night calm	W.
28	81	83	80	77	Fair. Very warm. At night calm *	W.
29	81	83	80	78	Fair. Vegetation withering, night calm	NW.
30	81	82	80	78	Fair, with light airs. At night calm	N.
31	81	83	80	80	Fair. At 3 p.m. a smart shower	W.

* At the Public buildings, and also at the Customs, the temperature at noon was 84°

AUG.	8 a.m.	Mid-day.	7 p.m.	10 p.m.	REMARKS.	WIND.
1	81	82	80	76	Heavy showers and thunder in the morning. 3 p.m. more rain	E. & N.
2	80	79	78	75	Cloudy and cool. Wind from various points	S NE & W
3	80	83	80	78	Fair	WSW.
4	81	85	83	82	Fair. Very warm. Evening dark clouds rapidly passing	W. by S.
5	82	83	82	81	Light showers in the morning. p.m. fair, with strong breeze	SW.
6	81	78	77	78	Raining until sunset	SW.
7	80	80	79	75	Fair. Night calm with heavy dew	N.
8	78	81	—	78	Fair. ditto ditto English mail arrived	SSE.
9	80	83	82	80	Fair. Flocks of "Curlew" reported to be passing south..	S.
10	81	84	82	81	Fair	S.
11	82	84	83	81	Fair. Three blue wing teal observed in Warwick Pond	S.
12	83	85	82	80	Fair	S.
13	83	84	82	80	Fair	S.
14	82	85	83	81	Fair. Pleasant breeze	SW.
15	84	85	83	—	Fair. ditto. A thunder storm passed to the northward	SW.
16	82	85	83	81	Fair. Observed a swallow in the town of Hamilton	WSW.
17	82	85	83	81	Fair. Three swallows hawking over White's Marsh	SSW.
18	83	86	83	81	Fair. Swallows abundant. Figs ripe	SW.
19	83	85	83	81	Fair. Fig trees shed their leaves	S.
20	82	85	83	81	Fair. The solitary sandpiper's note heard in White's Marsh. 6 p.m. distant thunder	SE.
21	83	83	82	79	Partially clouded, with thunder and very light showers	SSW.
22	75	83	82	79	a.m. fair. at noon showers	S.
23	81	84	82	80	Fair. at sunset dark clouds to the westward. 10 p.m. a few plover passing over	SW.
24	82	83	82	80	Fair with a light changeable breeze	V.
25	82	83	82	79	Fair. Temperature in Hamilton 84° to day	NW.
26	82	83	82	79	Fair, calm, and sultry. The nights for some weeks past have been mostly very calm	N.
27	82	83	82	79	Fair and calm. Vegetation struggling for existence, even the sweet potatoe looks nearly dead	NW.
28	80	83	82	79	Fair. A thin white mist prevailed both yesterday and this merning	NW.
29	81	85	82	80	Fair. A pleasant breeze until sunset	E. to S.
30	82	85	83	81	Fair. Vessels are moored in holes and corners to avoid hurricanes	SE.
31	82	85	82	—	Fair. A continuation of the same hot, dry weather	E.

This month was very warm, and at times oppressive, from the calmness of the atmosphere. Nothing could exceed the general stillness of the nights. Vegetation, except in low marshy grounds, was all but annihilated by the power of the sun's rays and the absence of rain.

SEPT.	8 a.m.	Mid-day.	7 p.m.	10 p.m.	REMARKS.	WIND.
1	83	83	80	78	Fair. Strong breeze. One red bird only sings in the garden	NE.
2	79	82	80	79	Fair, with strong breeze. Rice birds observed	NE.
3	80	83	82	80	Fair. Evening cloudy. At 9 p.m. a light shower	NE.
4	83	84	82	80	Fair. Blowing a fresh breeze	E.
5	80	85	83	82	Fair. English mail arrived. Killed rice birds in the marsh	E. & SE.
6	82	85	83	80	Fair. At night air from SW.	ESE.
7	82	83	82	79	Fair. Received a very fine bunch of white grapes from the Rev. J. Lightbourn	SE.
8	83	83	82	78	Fair. The sage bush shedding its leaves from continued drought	S.
9	83	84	82	81	Fair, calm, and sultry	S.
10	82	85	83	81	Fair. 3 p.m. therm. 85,5—these half degrees are not noted in these columns	S.
11	82	85	83	81	Showery, with strong breeze, scud passing rapidly. A tornado supposed to be raging to the westward	SW. & W.
12	83	84	82	78	Fair. The night clam and cloudless. 3 p.m. 84,5	WNW.
13	82	85	83	81	Fair. Evening cloudy. An owl seen	W.
14	81	84	82	80	Cloudy. Showers last night. Planted native seed potatoes for a Christmas crop	E.
15	82	84	82	78	Fair. Raining for some hours last night	W.
16	81	83	83	78	Sky partially clouded. Sultry and calm	V.
17	82	83	80	77	a.m. fair. p.m. threatening rain. Strong breeze	NE.
18	80	79	78	78	Cloudy. Evening luxuriously cool	E.
19	77	82	80	78	Fair	E. by S.
20	78	84	82	80	Fair. The Brig. 'Griffin' arrived from Halifax, N.S., under jury masts, having been totally dismasted in a hurricane on the 11th instant	S.
21	80	84	82	79	Gentle showers last night and this morning. Many long clouds floating in the sky	SW.
22	80	82	76	75	Beautiful showers in the night. Morning fair, with wind at W. At noon wind N. and at 2 p.m. NE. blowing freshly. Evening fair	V.
23	75	75	75	74	Cloudy, blowing fresh breeze, with light showers	NE.
24	74	75	75	75	Sky clouded, the upper stratum moving from the westward	NE.
25	75	78	76	76	Cloudy. With strong breeze. At night calm	E.
26	75	81	—	74	Fair. Sale of Alderney cattle	E.
27	76	79	76	74	Threatening rain. At 2 p.m. a thunder storm passed over Hamilton, with heavy rain at 9 p.m. therm. 73	E.
28	71	81	76	75	Dark all arround, with showers. Wind frequently varing. At night S.	V.
29	76	79	77	76	Fair. p.m. cloudy. At sunset heavy masses of white cloud over a deep lead coloured sky, extending from SW. to E.	W.
30	75	74	73	67	at 9 a.m. showers with heavy clouds and high wind from N. noon fair. Evening cloudy and blowing hard.	N.

OCT.	8 a.m.	Mid-day.	6 p.m.	10 p.m.	REMARKS.	WIND.
1	68	71	68	68	Cloudy, with strong breeze	N.
2	72	78	73	67	A beautiful day	W. to SW.
3	70	78	76	76	Fair. Governor F. Murray arrived, and "sworn into office"	SE.
4	74	81	78	76	Fair, warm and muggy. Snipes arrive from the north	S.
5	78	82	80	78	Fair. Sky somewhat clouded. 10 p.m. dark clouds and distant lightning, with rain in the night	S.
6	76	77	74	72	Cloudy, with strong breeze and passing showers	E. to NE
7	75	75	75	72	Cloudy. Heavy rain towards noon. Marshes flooded	E.
8	73	75	73	72	Continued showers, with fair weather between each	NE.
9	73	77	74	73	Fair. Several snipes shot in the Governor's Marsh	NE.
10	73	78	74	73	Fair	N.
11	75	76	76	73	Fair	N.
12	75	76	76	73	Sky partially clouded. Evening threatening rain	E.
13	74	77	76	73	Fair. General Assembly prorogued	NE.
14	75	—	74	68	Fair. Sky partially clouded	NW.
15	74	78	75	73	a m. Fair. 3 p.m. showery. Strong breeze	SW.
16	73	77	72	72	Fair. Evening cloudy, with scud flying from the west. Night clear	NNE.
17	73	79	75	76	Cloudy, with heavy showers. Strong breeze at night	S.
18	76	76	73	72	Cloudy. Showers in the early part of the day	NE.
19	73	75	67	69	Dark dismal day, raining continuously	E.
20	67	70	66	69	Dark and cloudy, with high wind and showers	NE.
21	66	68	66	72	Blowing hard, with heavy rain all the day. Much rain fell during this and the preceding night	NE.
22	67	68	66	67	Blowing a revolving gale all last night with rain, wind veering from SE. to SW. & NW. White's Marsh flooded. Observed many wild ducks there. 6 p.m. gale moderated. Bananna trees nearly all down	NW.
23	67	—	69	66	Fair, with a strong breeze. Many wild ducks shot	W.
24	68	78	71	68	Cloudy. Afternoon showery	SW.
25	73	80	72	70	Cloudy. Blowing hard, with showers. The brigantine "Penguin" on the rocks	SW.
26	73	81	75	71	Thick weather, with showers	E.
27	70	78	71	71	Fair. Heavy shower at midnight. Five couple of snipe shot in Pembroke Marsh	E,
28	72	78	71	70	Cloudy. Strong wind and driving showers. 6 p.m. rain	E,
29	72	78	71	71	Cloudy	SE,
30	73	80	71	71	5 a m. heavy rain. Cloudy. Four couple of snipe shot	SE,
31	69	80	70	66	A bright clear day. English mail arrived	W,

NOV.	8½ a.m.	Mid-day.	6 p.m.	10 p.m.	REMARKS.	WIND.
1	64	76	71	65	Fair. Heavy dew this morning	SW.
2	65	78	70	68	Fair. Warm and summerlike	W.
3	69	75	69	68	Fair. Chrysanthemums in blossom	NE.
4	70	75	68	64	Fair	NE.
5	72	72	68	62	Cloudy. At 10 p.m. blowing a fresh gale from the north	NE.
6	61	65	62	61	Cloudy. Blowing freshly, with light showers. Evening, wind moderated. Breath visible. Butter solid. Cockroaches disappeared. Children run round the garden to warm themselves	N.
7	62	70	62	64	Fair. A fire lighted in the mess-room at St. George's this morning	W.
8	64	70	65	65	Fair. Mignionette blooms	N.
9	66	75	66	65	Fair. Madeira onion seed sown in beds for general crop	NE.
10	65	72	66	66	Cloudy. Evening fair	E.
11	67	74	70	70	Cloudy. With a strong breeze. No snipes	E.
12	70	76	70	70	Fair. The white mulberry in blossom	SE.
13	68	77	70	68	Fair. The pride of India, and other deciduous trees, denuded of leaves	E.
14	65	77	65	—	Cloudy, with slight showers	E.
15	68	76	64	61	Dark, cloudy, dismal day, with rain. The s.w. wind howling through every aperture and crevice	SW.
16	66	65	64	62	Blowing a smart gale, with heavy showers	SW.
17	61	63	60	59	At day break, rain, hail, thunder and lightning, then cloudy, with strong breeze	NW.
18	63	64	—	65	Fair. White and red roses coming into bloom	S.
19	72	74	—	64	Fair. Blowing freshly. Showers at night	S.
20	60	66	60	60	Fair. Peach trees well in blossom	NW.
21	62	68	63	62	A dark gloomy day. Fire lighted in the evening	NW.
22	58	74	63	62	A beautiful day, p.m. wind s. English mail arrived	E. to W.
23	63	75	64	63	Sky very dark at times. Repeated showers	SW.
24	63	75	—	—	Bright and calm. Pic-nic at Mangrove Bay	N
25	67	74	67	67	Cloudy. Roses, chrysanthmums, & tournafouchia in blossom	SSW.
26	68	74	69	68	Fair, warm, and muggy	SW.
27	70	75	72	70	Fair, ditto ditto	S.
28	64	66	66	—	From an early hour in the morning to 3 p.m. continued rain, then clear, with light air from the w.	NE to W.
29	63	67	62	63	Fair	SW.
30	61	69	62	61	Fair. English mail arrived	W.

DEC.	8 a.m.	Mid-day.	6 p.m.	10 p.m.	REMARKS.	WIND.
1	62	75	58	54	Fair. A fire lighted in the evening. Temperature lower at night than for seven months past..	NW.
2	61	69	60	61	Fair. 4 p.m. rain with lightning & thunder. The troop ship "Resistance," from Quebec, with 26th Regiment on board..	W.
3	57	70	57	56	Cloudy	NE.
4	65	65	68	63	Cloudy. At noon a heavy squall of wind and rain, in which the wind veered from SW. to NW.	SW to NW
5	65	70	63	61	Blowing hard, with showers	S W
6	57	62	56	56	Cloudy, with strong breeze	N W
7	59	61	65	66	Cloudy, with showers. Night stormy	S W
8	63	63	62	57	Cloudy, with showers of rain	N W
9	57	62	59	57	Cloudy and cold	N.
10	62	68	62	—	Cloudy	E.
11	64	70	66	65	Fair. The kildeer plover met with in small flocks	SE. to S.
12	68	72	69	62	Fair, warm, and muggy. At 10 p.m. heavy rain	SW.
13	60	64	60	54	Fair. Scud passing from S.W. Head quarters of the 56th Regiment embarked for Cork ..	N.
14	50	64	53	54	Fair..	E.
15	64	69	66	67	Raining until 4 p.m.	E. to S.
16	68	69	66	65	Blowing a gale from S. At 3 p.m. gale moderated, the wind veering to SW...	S. to SW.
17	61	67	58	51	Fair. Troop ship "Resistance," sailed for Cork	NW.
18	51	—	60	63	Fair. Several vessels "*in distress,*" now in Hamilton Harbour	S.
19	63	72	65	66	Fair. Fine breeze	SW.
20	67	69	63	60	Dark and lowering. At noon rain and wind. Blowing freshly..	S. to W.
21	57	65	59	58	Cloudy	N. to W.
22	57	68	54	50	Cloudy and calm. p.m. light breeze. Night calm ..	W.
23	50	65	57	60	Cloudy. Rain at sunset. Garden strawberry in blossom..	N. by E.
24	60	66	57	58	Fair. Night calm. Severe colds prevailing	E.
25	55	69	61	59	Fair. A red bird sings at early dawn	WNW.
26	62	68	62	62	Showers. Patches of onions already planted out ..	N.
27	56	68	62	62	Splendid day. Freemasons walked in procession ..	S.
28	62	71	—	64	Showery, with strong breeze. English mail arrived *	S. to SW.
29	63	72	—	67	Fair. Lauristinus in blossom	SW.
30	62	72	62	61	Blowing a smart gale, with heavy showers and one flash of lightning. Strong wind all night	S. to W.
31	56	62	58	56	Cloudy with strong breeze. End the year with a blazing fire.	NW.

* On this day the writer received orders from the Treasury to make over the duties, books, papers, etc. of his department, to the Island Authorities, on the 5th of the following month; and from that date has been on a "compensation allowance," by no means calculated to flatter the feelings of an officer, who, for upwards of twenty-four years has devoted his energies to the service of his country abroad. J.L.H.

UNUSUAL FALL OF RAIN.—On the night of the 7th of June, 1850, and until daylight on the following morning, the rain descended in a tropical deluge upon the Bermudas. This heavy rain was accompanied with distant thunder and lightning. White's Marsh, near my residence, was flooded with water, and Mr. W. B. Smith, the Receiver-General of the Islands, informed me, that on the same night, a large tank, which he built under his residence at Riddle's Bay five and twenty years before, overflowed with water for the first time since its construction. Previous to this event the catch of water had never reached within some feet of the upper margin of the tank. As another proof of the unusual quantity of rain which fell on that occasion, I was told, on excellent authority, that twenty-five tons of water were obtained on board the "Tenedos" convict hulk that night, and all from the catch of her own roof.—J. L. H.

FRESH WATER SOURCES.—It is not uncommon to hear the native inhabitants of the Bermudas speak of their "Springs of Fresh Water." Now, this is merely a *facon de parler*, for in reality there is not the smallest rill of running water in those islands; the so-called "springs" being nothing more than wells sunk a few feet in depth on the lower levels, particularly along the edge of the north-shore. From these, a supply of water is obtained which can be used for ordinary purposes, though the water in them is subject to rise and fall with the action of the tides.*

There can be little doubt that these wells are supplied

---

* It is true that small underground rills of water are sometimes met with in draining the marsh lands near the level of the sea, towards which they invariably run.

by rain water percolating through the porous conglomerate of shell of which the entire Islands are composed, and that such rain water is met at a certain level by the water of the ocean penetrating through fissures or forcing its way by means of filtration. That fresh water, from being specifically lighter than sea water, will, when undisturbed, float for some time upon the denser fluid of the latter, without much admixture of the two, is highly probable, but the bare contact of rain and sea water, in any degree, must, I apprehend, tend to deprive the former of its purity. Under such circumstances it is difficult to conceive how well-water should be fit even for the use of cattle.

I will cite a case in support of this opinion.

On my arrival in Hamilton, the principal town of Bermuda—now some years ago—I observed that a well, a few yards distant from the Steam Grist Mill, was much resorted to for water, and on my enquiring if it were fit to drink, I was assured there was no finer water in the Bermudas. A similar well had been sunk inside the engine house of the said mill, for the express purpose of supplying the boilers (high pressure of ten horse power) with "fresh water." The mill was worked with the water obtained from this well for five or six years, when a slight explosion of steam caused the boilers to be opened, and the interior of each was found to be coated on the bottom and sides with a solid crust of indurated grey salt and calcareous sediment about two inches thick, which required a stout blow from a sharp iron pick to detach even a small portion of it; thus proving, beyond a doubt, that water, hitherto considered to be *fresh*, was in reality impregnated with a portion of *salt*.

To show the influence this water has upon cattle, I am induced to insert the following from my note book :—

"March 31st, 1851.—Mr. Somers Tucker, the Government beef contractor, who has upwards of a hundred bullocks in his stalls, and has been out of tank-water for some weeks, tells me that he is compelled to make use of the water from the Corporation well, in front of Dean's school, for watering his cattle; that he drains the well for that purpose every day, and gives them as much as they please to drink. He finds the well water scour the cattle to an injurious degree, and has endeavoured to counteract the evil by lessening their allowance of Indian corn meal, and increasing that of upland hay.

"June, 1851.—Mr. Tichener Darrell, of Warwick Pond, tells me that he keeps three or four cows, the produce of which, in the shape of fresh butter, is sent to the Dockyard for sale, were it is well known for its superior quality. He gives these cows tank water only, and assures me that if a cow be allowed to drink the common well or moat-water, as he termed it, the butter produced from the milk will absolutely 'stink.' For this reason every care is taken to keep the cows from touching it."

During the scarcity of rain-water in the drought of 1849, many of the poorer inhabitants were reduced to the necessity of using well-water only, which had a most injurious effect on the health of those persons, and was rapidly causing the spread of diarrhœa and fever, when a fall of rain happily replenished the many exhausted tanks, and rapidly restored the sufferers to their usual state of health.

WATERSPOUTS.—September 12th, 1845. Hearing that a waterspout was to be seen in the vicinity, I ran out, and for the first time in my life witnessed this phenomenon.

To the eastward of the Court-house Hill, on which I stood, appeared a dark cloud, slowly moving from north to south, and, apparently, about one mile distant, from a portion of which rain was falling.

From the centre of this cloud, and in a part of it free from the falling rain, descended a magnificent column of the same colour as the vapour above. This column was straight and almost perpendicular, and was pronounced by a spectator of respectability, who had resided thirty years in the Bermudas, to be the finest waterspout he had ever beheld.

I much regret that an intervening hill prevented the base of this column from being visible to me, as it must have passed almost in immediate contact with the southern shore of the Islands.

This column of vapour—for it could be nothing else—appeared to move at the same rate and in the same direction as the cloud, revolving in a direction the reverse to that of the hands of a watch. The inside or centre of the column appeared to be hollow, and filled with rapidly ascending vapour, closely resembling the passage of smoke up a chimney. After an existence of some minutes, the column began to lose much of its density, particularly at the extremities, and eventually the central portion of it was alone visible, suspended in mid air, and still revolving. Much gratified with what I had seen, I was on the point of leaving, when a second waterspout was observed to be forming on a portion of the same cloud, not far distant from the preceding one.

This at first appeared suspended from the lower edge of the cloud, in the form of a long pointed icicle, widened at the base, which gradually extended and enlarged itself in a downward direction, until it exhibited a grand column of vapour, in every respect like the former, and revolving in like manner. During the latter half of its existence it assumed a graceful curve in lieu of its perpendicular appearance, then gradually disappeared at the extremities, leaving, as before, a whirling centre to disperse slowly in the surrounding atmosphere.

I was surprised to find so little resemblance to anything like a volume of water in the appearance of these so-called waterspouts; indeed, with the exception of a shower of rain which fell from another part of the cloud, I saw nothing to induce the belief that water, in the fluid state, was present during the phenomenon. On the contrary, the scene I had witnessed left an impression on my mind, that these waterspouts can be nothing more than the dense vapour of a cloud whirled into the form of a hollow tube by the force of some revolving squall of wind, thus filling the vacuum or centre of such revolving atmosphere to the fulness of its height and depth, as it sweeps over the surface of the earth.—J. L. H.

AURORA BOREALIS.—At seven o'clock on Monday evening, the 29th of September, 1851, the heavens became beautifully illuminated by the Aurora Borealis, which extended over the crest of the intervening hills of the north shore, far to the east and west. The lower portion visible from my residence, exhibited the usual white light of the Aurora, more or less vividly at intervals, and was bounded by a wide and luminous arc of the most beautiful roseate or carmine, extending almost to the altitude of the Polar star, and ultimately much above it. Through this reddened portion of the sky, evanescent rays of white light were continually shooting upwards. At nine o'clock the brilliancy of the phenomenon had passed away, and at ten, when I last saw it, the northern part of the heavens was more faintly illuminated.

The night was calm and fair, with a light air from the south-east, and a young moon, which set between eight and nine.

This is the second time that I have observed the Aurora Borealis in the Bermudas, during a residence of nearly eleven years. On the former occasion it appeared on the 17th of November, 1848. Familiar as I have been with this phenomenon in the latitude of 46° of the North American colonies, I never before observed the heavens radiant with the beautiful deep rose colour which prevailed on both these occasions.

November 2nd, 1851. In a London newspaper of the 4th of October, 1851*, which arrived by this day's mail, it is stated, that at twelve p.m., on the evening of the 29th of September last, the Aurora Borealis was very vivid in London, and that its brilliancy at one time was supposed to arise from some tremendous conflagration in that metropolis.

Allowing for difference of longitude, the appearance of the Aurora in the Bermudas and at Halifax, Nova Scotia, was simultaneous with that recorded in London, and the fact of its being visible at the same moment from parts of the

---

* Vide Bell's Weekly Messenger.

globe three thousand miles distant from each other, is an interesting proof that it was far beyond the supposed limits of the terrestial atmosphere.—J. L. H.

FROST.—On the evening of the 24th of December, 1840, the wind being northerly, and the sky bright and clear, an unusual degree of cold was experienced in the Bermudas. At this time I was a stranger in the Islands, and had been led to believe that the winter season was so mild and agreeable, as to render fires unnecessary. Moreover, I was residing in a house, the walls of which were green and purple from the damp; where the staircase from the hall below, entered the drawing room, without the intervention of any doorway, and where the only fire-place was in the kitchen. I can safely say that I suffered more, on this occasion from cold, than during the whole of my long rèsidence in a Canadian climate.

On the following morning white frost was visible in low situations; water in tubs was frozen to the thickness of half a crown; and the subsequent blackened appearance of young lettuces and potatoe stalks, sufficiently evinced the effect of this visitation upon the vegetable kingdom.

Of course, "the oldest inhabitant" was quoted as never recollecting such an instance of frost in the Bermudas; but, if we may be allowed to judge of past events by those which happen in our own time, I should feel disposed to question the memory of this much-respected and universal character; indeed, I have been credibly informed by persons resident in Ireland Island, that one or two instances of frost have been observed at that station within the last few years.—J. L. H.

The following paper on "Experiments made at Bermuda on the Carbonic Acid in the Atmosphere, in the year 1836," from the pen of Lieut.-Col. Emmett, appeared in the London and Edinburgh Philosophical Magazine of 1837.

PER QUANTITY.

EXPERIMENT I.—September 25th. A glass receiver of 3920 cubic inches, = 15·5 gallons, was taken to the north side of the island, beyond any building. Wind north, day fine, thermometer 79°. Into this, after well washing with rain water and collecting the air, were put 1500 grain measures of lime water. The receiver was then well closed with a cork and set aside.

September 24th, 4½ p.m., thermometer 82°. Some of the lime water used was tested, 1500 lime water taking 410 test sulphuric acid, the liquid would be 1·009 test. It took 330 to saturate the remaining lime water, consequently left 80 for the carbonic acid in the air. (The receiver was not long enough exposed; my bottle was two gallons; it was exposed three or four days, and agitated to exhaust the air. Consequently ten times as much would probably be required by ten times the size of the bottle.)

EXPERIMENT II.—September 25th. Receiver and acid as before, and the lime waters but 210 for neutralization; wind strong from south-west; thermometer 80°.

September 28th. The lime waters from the receivers took 120 grain measures for neutralization, leaving 90 for carbonic acid gas; a very nearly similar result as before.

EXPERIMENT III.—October 2nd; at 5 p.m.; wind south-west, at the cessation of a heavy gale, with much rain; thermometer 78°, barometer 30·00. Receiver and acid as before. Lime water required 390 measures for neutralization. Tested that in the receivers at 5 p.m. of the 8th instant; thermometer 75°. This required 210 measures of the acid for neutralization, leaving 180 for carbonic acid, being double that before, or about 1 in 3920.

EXPERIMENT IV.—October 11th; 4½ p.m. Collected air as before; there had been much rain during the day, but it was fine and calm after 3 p.m.; thermometer 77½°. In this case the lime water took 375 measures for neutralization. Tested that in the receivers on the 18th; thermometer 75°. The 1500 grains in the receiver required 280 grains for neutralization, leaving 95 for carbonic acid gas.

In the experiments 1, 2 and 4, the gas is consequently about 1 in 8000; and in the 3rd 2 parts in 8000. In the 3rd the receiver was out during the rain, but so placed as to prevent its entrance. The air in 2 and 4 had traversed the small island of St. David's, distant perhaps 1¼ of a mile, thinly inhabited, and thence the inlet of the sea, St. George's Harbour.

The acid was pure, brought out with me for particular experiments.

Looking to the general result, and in number 3, the quantity being *double,* inaccuracy of observation of the measures might possibly have led to the differences.

MISCELLANEOUS NOTES, ETC.

AIR-BLOWN FISH.—It is the custom of Bermuda fishermen, when they take large fish, such as Groupers and Rock fish, to pass a small rope through their gills and tow them alive into port. The fish thus brought to market are almost always so blown with air as to float belly upwards, being unable to retain their natural position in the water. The fishermen assure me this does not arise from towing, and that all fish caught in *deep water* invariably come to the surface in that state. Shallow water fish, they say, are never blown in this manner.

Again, the Porcupine fish (*Tetraodon*) when hooked, comes to the surface expanded with air into the form of a ball, with its formidable spines erect; but when taken in a fish-pot it does not exhibit this extraordinary inflation, and the spines are recumbent.

How are these phenomenon to be accounted for?—J. L. H.

MORTALITY AMONG IMPORTED POULTRY.—It is an extraordinary circumstance that common domestic fowls, imported from the United States of America into the Bermudas, though landed in the best of order and condition, sicken and die soon after their arrival in the Islands; whether from change of climate or food, or from what other cause I cannot say, but certain it is, that in the course of a few weeks, dozens of these birds will dwindle down to as many units, notwithstanding every care has been bestowed upon them. Geese, ducks, and turkeys appear to be exempt from the cause of this mortality; and fowls bred in the islands do not suffer from it, even when associated with the imported birds. I have occasionally attempted to improve the few fowls about my own residence by purchasing handsome birds from American vessels, but after moping about for some time with drooping wings and pallid combs, they invariably died.—J. L. H.

North American Indians.—August 19th, 1850. A coloured man, native of the Bermudas, named "Nat Kiel," resident on the verge of the rolling "Sand Hills" of Paget Parish, called on me a few days ago, and stated that he captured the Scarlet Tanager which Mr. Marriott sent to me on the 19th of April last. That he first observed it among the peach trees which grow in front of his small wooden house, and caught it in a trap of horse hair nooses set for cat-birds on the rim of a tub of water; that after keeping the bird in confinement for three days, it died. He saw no other Scarlet Tanagers at the time, nor did he ever previously observe the bird in these islands.

This man, Nat Kiel, has so thoroughly the features, lank hair, and full black eye of the Mic Mac Indians of North America, and is, moreover, so perfectly distinct from the different races of African and Asiatic blood prevalent in the Bermudas, that I think there can be little doubt of his Indian origin.

Robertson says, in his History of America, that many Indians of the Mic Mac tribe were captured by the settlers of James's River, in the early settlement of that portion of the coast. These were sold and exported as slaves, many being sent to the Bermudas.

Mr. W. B. Smith, the Receiver-General of the islands, tells me that in his boyhood he has often heard his father speak of certain families being descendants of North American Indians, and that "Nat Kiel,"—now about sixty years of age—has always been considered a person of this description.—J. L. H.

BLANK SEASONS.—October 25th, 1850. I have assiduously hunted the neighbouring marshes, cedar groves, and sea coast, up to the present period, and regret to record that the migratory season has nearly passed away without any influx of the feathered tribes from the north, or any other point of the compass; indeed, it promises, like some former ones, to be almost blank and devoid of interest to the ornithologist. Not a single addition has been made to our list since the 12th June, though my ornithological friends have been keenly on the *qui vive* for novelties.

THE FLAMINGO.—July 1st, 1850. The "Peri," a small vessel belonging to the Bermudas, and recently returned from an onion voyage, brought from the island of Inagua, (one of the Bahamas,) two young flamingos. The master, J. T. Yates, tells me that he had about a dozen of these birds on board, all of which died at sea, excepting the two here mentioned. They were the property of passengers, and were captured the day before the "Peri" sailed, by a party who walked into the country for that purpose.

July 4th, 1850. Met with one of the "Peri's" passengers, an intelligent native named "Hollis," who had been present at the flamingo hunt above mentioned, and questioned him regarding what he saw of the habits, etc., of that bird. He stated that with some companions he visited Lake Rosa, at a distance of fifteen miles from the port where the "Peri" lay, and waded to some of the islands which dotted its surface, the water being only knee deep. On one of these islands they disturbed a large number of flamingos, at least two hundred, which were too shy to admit of a near approach, and were all in red plumage; these would settle on some distant margin of the lake in line, "resembling a company of soldiers." On reaching the rocky shore of the island in question, many young flamingos were discovered, some of which were run down and captured. They have an awkward gait, but scuttle along at a good pace. These were in the grey plumage, and of different stages of growth; the larger, just putting forth the quill feathers of the wings. Hollis confidently states, that he saw upwards of a thousand old flamingos on the lake that day, or rather on the small portion of it visited by him. He also saw many nests of these birds, and found several of their eggs, which appeared to have been thrown out by the parent birds, and proved to be addled. They were white, and about the size of a common goose egg. The nests were composed of mud and sticks, more or less raised on account of surrounding water; the highest of these were certainly not more than nine inches from the ground, while many others were nearly level with it. The surface was hollowed out, and capable of containing about two eggs, not more. I referred to Wilson's "American Ornithology," and read the paragraph which describes the elevated nest constructed by this bird, to admit of its long legs dangling on each side during the duties of incubation; at this, my informant smiled, and assured me that he saw nothing of the kind; that he had particularly noticed many of their nests, and that in no one instance did the height of any of them exceed what he had already stated.—J. L. H.

Sea Bottle.—May, 1849. That very curious marine plant, commonly designated the "sea bottle," has been common on the sands during the whole of this month. These "sea bottles" are transparent, and shaped like a small balloon; the largest measuring about one inch in length. The inside is filled with sea water, sometimes perfectly clear, at others deeply tinged with green, arising from small particles of sea weed which have become hermetically sealed up with the confined water. I have sometimes found these bottles washed on shore in small clusters, as if they had sprung from the surface of some submerged rock. They are without leaves or branches.—J. L. H.

Mushroom.—March 2nd, 1852. Mr. Barss came to me this morning with a fine specimen of the mushroom, (*Agaricus campestris,*) in his hand, which had been gathered by Mr. Thomas Hall upon his own ground. Though resident in the Bermudas upwards of eleven years, I was not aware, until now, that the mushroom was known here. Mr. Hall tells me he has noticed them for many years at this season, growing wild on the same spot,* and that he has seen as many as a dozen at one time, some of them as wide as the palm of his hand.

Might not the cultivation of this fungus be turned to good account in a winter gardening climate like this? Asparagus, one would suppose, would also thrive here, and yet I have never met with it.—J. L. H.

---

* A grassy bank between the east wall of Pembroke Churchyard and the adjoining marsh.

SMOKE.—In June, 1849, the wind prevailed much from the north and north-west, and for a whole week, viz., from the 12th to the 18th of that month, the Bermudas were enveloped in a white mist, not unlike London smoke, and which was found by the Royal Mail steam-packet to extend from the port of New York to those islands. This mist deadened and obscured the rays of the sun, causing it to shine with greatly diminished power. Many persons considered the mist to be sea fog, while others pronounced it to be smoke.

Subsequent intelligence proved it to be smoke, arising from fires which had been raging for some weeks in the forests of New Brunswick, Nova Scotia, and Prince Edward Island, and which the prevalence of northerly winds had driven across the ocean to the Bermudas, a distance of full seven or eight hundred miles.

These fires were all induced by the extreme dryness of the season in the British North American Colonies.

The following extract from the "Prince Edward Island Gazette," of June 19th, 1849, bears upon the subject:—

"Fires have been raging in the woods in all directions for the last fortnight, and have caused serious apprehensions that, not only the wood would be destroyed, but that fences, houses, barns, and even crops would be swept off by the devouring element; the atmosphere being surcharged with smoke to an unprecedented degree. We regret to say, that rumours have reached us that, in some localities, these apprehensions have been, to a degree, realized. In the neighbourhood of Bedeque, Crapaud, New Devises, Johnston's River, and Monaghan Road, it is said that much damage has been done; but as the people are occupied night and day watching and endeavouring to suppress it, we cannot obtain correct particulars. Large fires have been seen on the mountains of Nova Scotia, and from the volumes of smoke which have been sent over the Straits, we have reason to think, must have been productive of much damage. New Brunswick, especially in the neighbourhood of Miramichi, we hear, is also suffering from the same calamity."—J. L. H.

Discovery of Ancient Hawse-pipes.—In or about the year 1844, I was informed by Mr. George Somers Tucker, of Hamilton, that within the preceding two or three years, a native settler, while fishing within a short distance of the "Spanish Rock," on the south shore, got up from the bottom, the leaden hawse-pipe of a vessel, which had evidently been a great number of years under water; and soon afterwards, the same party fished up the fellow to it, in the same place. One of these hawse-pipes was cut to pieces by the finder, and converted into sinkers for nets and lines; the other was purchased by Mr. Somers Tucker, and conveyed to Hamilton, where it was deposited in his grain and flour store. This hawse-pipe he promised I should see; observing, that he intended to send it home to the British Museum as a curiosity. Search was accordingly made for the same, again and again, but all to no purpose, compelling its owner to conclude that it had been stolen from the store, and disposed of as old lead.

I know not at what period of history leaden hawse-pipes were used by the ships of European nations, but if the above statement be correct—as I have every reason to believe it to be—there certainly appears to be ground for supposing that some unknown vessel was wrecked at this very spot (sixty-six years prior to the settlement of the Islands), and that the inscription on "Spanish," or "Spaniard's Rock," has been cut by some of the hapless survivors of its crew, correctly shewing the date of the catastrophe.—J. L. H.

THE EARTHQUAKE.—On Tuesday night last (March 2, 1858), a sensation, happily but little known in these Islands, was felt from one extremity to the other of the group. As soon as the surprise of the moment allowed time for reflection, all perceived that Bermuda was being shaken by an earthquake, and many, especially those who elsewhere have witnessed the dread effects of this disastrous phenomenon, became much alarmed. Many hastily prepared themselves for escape, and others began to cast about for some place of security and safety.

The shock occurred as nearly as possible at thirty minutes past ten o'clock, and continued for several seconds, how many it is extremely difficult to conjecture, but all, the experienced as well as the novices, observed that it was a very protracted one. We have heard several West Indians say that they have seldom felt a more decided or more prolonged shock.

The direction in which the earthquake travelled seemed to be from South West to North East, or from South to North. An accurate statement, however, on this point can only be obtained from a comparison of the time when it touched the various parts of the Islands, and this cannot easily be obtained, when there is no common public standard of time.

Those who saw the water in Hamilton Harbour say that it was much and peculiarly agitated, as is so frequently the case on such occasions. Vessels strained hard at their anchors, and persons afloat were a good deal tossed about. The shock was far more severely felt by those who were lying down, than by those who were walking or sitting. Children jumped out of bed in great alarm, and even the uninitiated at once exclaimed "it's an earthquake."

The noise which accompanied, or as some say preceded the shock, was a hollow rolling sound, not unlike distant rattling thunder, but longer and more regular in tone. Our own opinion is, that this noise was only heard towards the end of the shock, but others say that it was the commencement of it.

We have not heard that any damage was done.

One gentleman who felt the shock so much that he was unwell for several hours, thinks that there was a second, about 2 a.m. of the 3rd. Some others speak of having felt the second shock. We have heard of three or four persons who suffered from nausea after the shock.

The "Ocean Bird" was at sea about thirty miles off, the Light House bearing N.W., at half-past ten on the night in question. Captain Peniston reports having at that hour felt the earthquake. The "Ocean Bird" was running about eleven knots, when suddenly those on board experienced an alarming shock, which led them to suppose they had struck a rock, and that there was some error in their calculations. Captain Peniston says that just before, he saw a heavy cloud rise in the S.W., from which he expected a heavy squall and rain, but it passed away with the shock, but little rain falling. The following day it was observed that the water was very muddy.

At Mount Langton the shock was very perceptible, and

His Excellency the Governor judges that it must have lasted at least one minute.

The shock was more severely felt, as far as we can judge, in houses built on solid rock than in those having their foundations in soft stone or soil. The latter of course would offer less resistance to the violent commotion of the earth, and would yield readily to the pressure. But whether, as many suppose, the more elevated parts of the Islands were more severely visited than the less, is not so easy to determine. Many think that the hills shook much more than the valleys.

We may, we fear, expect to hear that this earthquake was widely felt. It is not very unusual for slight shocks to be felt here; but it is many years since one similar to that of this night week was experienced in Bermuda. Generally, the effect is so trivial as to escape the observation of all but an experienced few; very different was that of the 2d inst.

If the Colony at large will remember how recently some of the fairest countries of Europe have been desolated by the visitations of these dreadful phenomena, it can hardly fail to acknowledge how much it owes to the goodness of God for sparing Bermuda those awful results which elsewhere have attended them. To feel the solid earth rock and tremble under one's feet as if its huge vertebræ were suddenly thrown out of joint, is a strange sensation, well calculated to overawe the human mind, but it is perhaps yet more amazing that such mighty commotions can come and go without leaving a trace behind—not one heap of stones or one riven wall to mark their progress.—*Bermuda Royal Gazette.*

Sharks of a large size are numerous on the outer reefs of the Bermudas, and though specimens are occasionally killed,

I was never so fortunate as to examine one (the mackerel porbeagle excepted), and therefore remain in ignorance as to the species. I have a tooth in my possession taken from a large shark, which, with many others, followed a dead whale into a small bay on the south shore, where the whalers towed it for the purpose of *flinching.* When attacked with harpoons, this shark seized the bottom of the boat with its capacious jaws, and left two or three of its teeth in the woodwork. The jaws were presented to the Bermuda Museum, but having been improperly prepared, became so offensive as to cause them to be thrown away. The teeth were *as sharp as lancets*, literally so, when first brought in.

On the 1st June, 1849, a sperm whale drifted ashore to the south of the Light-house, supposed to have been killed and lost by the crew of an American whaler then cruizing off the Islands. One side of this whale was almost entirely consumed by sharks, there being about fifty of these voracious animals about the carcass. One of them was killed, and measured seven feet in length; two buckets full of blubber, and a portion of a green turtle were taken from its stomach.

Again, on the 27th September following, a coloured man called at my house with the backbone of a large shark for sale. He stated that the shark was killed "away in the deeps" from a small row-boat, a few days previously; not with hook and line, but by running a noose over its tail, and towing the animal to the shore, when it was found to have "drowned itself." My informant assured me that it measured nine feet in length, and that he sold the jaws, containing six rows of teeth, to an officer at the dockyard. Judging from the vertebræ I saw, I do not think the length exaggerated. From the mode of its capture might we not presume this to be the Basking Shark (*Selachus maximus*)?

On the 15th January, 1851, a shark eleven feet in length was captured with a hook and line at Ireland Island, and when opened, was found to contain the head and fore quarter of a calf with the skin upon them, also two turtles with shells about nine inches in length, and a few other delicacies. One of the turtles appeared to have been killed but a short time; the heads of both had disappeared. The species of this shark was not ascertained.—J. L. H.

YELLOW FEVER.—It is customary with many of the medical profession in the Islands of Bermuda, to describe yellow fever as an epidemic, generated by local causes existing in the atmosphere, which local causes are ascribed to the miasma of swamps, ill ventilated and dirty streets, crowded convict hulks, and so forth. Then the disease is stated by them to be neither infectious or contagious in its nature; and in support of these views scientific terms and professional statistics are largely appealed to.

I am quite aware that the opinion of a mere non-professional individual can have little or no weight in the estimation of these gentlemen. Notwithstanding this discouragement, let us enquire into the truth of these assertions, and endeavour to approach the real character of the disease.

With regard to its being generated by local causes—what is there in the shape of marshes, pools, etc., that can generate a deadly miasma? The marsh land of the Bermudas is trifling in extent, and from my own experience I can truly affirm, that the proximity of a marsh is neither unpleasant to the senses or prejudicial to health. Of stagnant pools, there are a few which give forth unpleasant exhalations during the night, but these vapours do not appear to affect the health of the cottagers who reside in the immediate vicinity. The ponds near Paget Church may be cited as an example in point. One glance at the map of Bermuda will be sufficient to convince any reasonable person, that an island little more than a mile in width, isolated in the vast Atlantic, and distant six hundred miles from the nearest point of the American coast, can have no atmosphere of its own—swept continually by the winds of the ocean, malaria can no more exist in the Bermudas than on board a ship at sea. If the narrow and dirty streets of St. George, or the convict hulk there stationed, could generate disease, why was that locality free from yellow fever during the twenty years preceding the attack of 1843 ?* The causes of fever cannot, then, with reason be said to exist in the sea air breathed by the inhabitants of these Islands.

Now let us consider the infectious or contagious nature of this fever—these are terms which I consider to be synonomous, and applicable to any disease which may be taken by approach or contact with the sick—all quibbling on the derivation of the word "contagion" notwithstanding.

If the statement of certain members of the faculty be correct, viz.—that yellow fever is not infectious ; why, let me enquire, are ships of war with this fever on board, not allowed to send their sick to the Royal Naval Hospital at Ireland Island, but either have to ride at anchor in the offing with the yellow flag flying, or proceed to Ports Island, in the sound, as a fever and quarantine station ; and why is all communication prohibited with such ships and stations?

* The same observation will apply to other parts of the Bermudas.

The precautions adopted by the authorities on the arrival of the French frigate L'Armide, in 1852, were in accordance with this practice. Again, if yellow fever be not infectious, why is the garrison scattered in detachments, and placed under canvas in retired rural spots, when attacked by that disease? and why was the practice adopted, in 1853, of burning the bedding of all military patients who died of fever? Here, then, we have a portion of the medical profession gravely asserting that yellow fever is not infectious, and by a strange professional inconsistency, acting in direct opposition to those views, thereby destroying in a great measure, all public confidence in their proceedings.

If yellow fever be not endemic or indigenous to the Islands of Bermuda, how are we to account for its appearance there at distant and uncertain periods? It prevailed in 1780, in 1818 and 1819, in 1837 at the dockyard only, in 1843 and in 1853. On every one of these occasions yellow fever was fearfully destructive in the West India Islands, and it was generally believed that the disease was imported from those parts. There is certainly strong evidence to bear out these statements, and my own personal experience of the fever which visited the Bermudas in the years 1843 and 1853, leads me to the belief that such has always been, and ever will be the case, while intercourse by shipping is carried on with that part of the globe. Steam navigation, and the difficulty of imposing sufficient quarantine regulations on steam ships, must greatly add to the frequency of such results; of the efficiency of quarantine, the case of L'Armide is a triumphant example.

To bring the origin of fever nearer home, let us look closely into the nature of that which prevailed in the Bermudas in

1853. No one will deny that it differed materially from former fevers, by its peculiar malignancy and destructive powers—by symptoms of the disease unknown to medical officers who had been accustomed to treat the ordinary yellow fever of the West Indies. It is notorious to all the world, that such was the character of the fever introduced into the Brazils by a slave ship, in 1851, to the destruction of seventy thousand of its inhabitants in that year—that such was the character of the fever which spread from thence to the French, Dutch, and British Settlements in Guiana, and from them to the West Indian Archipelago in 1852, and again in 1853.* How is this remarkable identity of character to be accounted for, unless we admit the actual importation of the disease, and the principle of infection?

Hundreds of cases might be quoted to prove that yellow fever is a highly infectious disease. Gibbon, the great historian, states, that the plague was transferred from infected persons, to the lungs of those that approached them, by mutual respiration, and I am disposed to believe that yellow fever extends its poisonous influence much in the same manner, notwithstanding the many statements to the contrary.

It is much to be feared, that by acting upon the principle of non-infection, in cases of yellow fever, many valuable lives have been sacrificed, which timely precautions might have saved. How much wiser would it be, when this deadly malady prevails, to err—if human nature must err—on the safe side of this question.

* It also visited New Orleans with frightful severity in 1853.

RETURN OF DEATHS AMONG THE TROOPS OF THE BERMUDAS, DURING THE FEVER OF 1853.

Officers.	Officers' Wives.	Officers' Children.	Men.	Women.	Children.	Remarks.
15	4	6	297	25	24	Total, 371.

The above return I obtained from the Fort Adjutant's Office, on the 6th January, 1854. J.L.H.

CURIOUS CIRCUMSTANCE.—July 20, 1852. A Bristol Barque, called the "Camœna," bound from Jamaica to London, with a cargo of sugar, rum, and logwood, was unfortunately stranded on the reefs of Bermuda, about the beginning of last month. A portion of her cargo, together with the rigging and materials of the ship, were saved by the island boats, and brought into the port of Hamilton. The hull, with the remainder of the ship's cargo, consisting of two hundred hogsheads of sugar and some logwood, *all under water in the hold*, was sold at auction for the sum of fifty-one pounds.

The enterprising purchasers soon discovered that a considerable portion of each hogshead of sugar had sustained little or no injury from the salt water; and this sugar for a whole week, they were busily engaged in saving. Curiosity led me to examine this shipwrecked sugar. I found it laying in bulk, in a warehouse at Hamilton, to the amount of one hundred barrels, or perhaps more; it was not wet, nor could the slightest indication of the presence of salt be detected by the sense of taste. Indeed, I very much doubt if any person, unacquainted with its history, could have distinguished this sugar from what was saved from the same vessel, without having been in contact with the water.

May we not account for this occurrence by supposing the salt water in the hold of the vessel to have been so thoroughly saturated with sugar, as to render it incapable of holding more in solution, and from the superior gravity of this saturated water, and the probable small amount of leakage in the ship's bottom, little disturbance could take place with the external waters of the ocean? In no other way can I understand how so large a portion of sugar could remain immersed in sea water for many days without apparent injury.—J. L. H.

THE SAND HILLS.—On the south shore of the Bermudas, about midway between their east and western extremities, and not far from Paget Church, is to be found an extensive drift of comminuted shell, which from its resemblance to ordinary sea sand, is generally termed the "Sand Hills."

This accumulation of drift shell extends along the shore for upwards of half a mile, and advances inland for nearly the same distance. It appears to be of very recent date. In every south and southerly breeze, the driven shell may be observed moving briskly over the gradually ascending surface of this drift, and gently rolling down the elevated, and even slopes of its extreme inland termination, at an angle of about forty-five degrees, filling up the undulations of the surface of terra firma, and burying in its onward progress cedar groves and cottages. The accumulation is rapidly increasing, and, viewed from the higher ground near Hamilton, is seen towering above the steeple of Paget's Church.

In high winds the driven shell is carried to a considerable distance beyond the accumulated drift, and then resembles a haze or mist overhanging that portion of the neighbourhood.

When we consider the peculiar formation of the Bermuda Hills—the absence of all other material in their geological structure—from the vast beds of loose comminuted shell, interspersed with small sea shells entire, to the so-called limestone rock, varying in its solidity from " soft" to "hard," and the dip or cleavage of the same, may we not infer that the entire group of islands has been formed precisely as the Sand Hills of Paget Parish ?

The circumstance of finding the skeleton of a duck with two fossil eggs under it, embedded several feet below the surface, in " solid rock," which was removed several years ago to level the surface of the present Dockyard, may tend to dispel any doubt which may exist upon the subject. This natural curiosity was seen by the late Rev. Mr. Mantach, in the possession of an officer of the Royal Engineers, by whom it was carried to England.

Before concluding these brief observations on the wonderful and comparatively recent formation of the Bermudas, from the depths of the Atlantic, and upon the forms of animal and vegetable life there found to exist, let us look forth upon the wide belt of ocean which surrounds the group, and enquire how far the winds and waves have incidentally contributed to clothe those islands with the trees and plants found upon them by early navigators and settlers.

During the summer months the winds chiefly prevail from

the south, south-west, and west, and in the winter season from west to north-west and north ; easterly winds are much less frequent.

The warm water of the gulf stream, impelled by westerly winds, occasionally deviates from its ordinary course, and washes the rocky coast of the Bermudas. Gulf weed (*Fucus notans*), is never absent from the creeks and bays, where it accumulates in sufficient quantity to be worth collecting for purposes of manure. Logs of squared pine timber, covered with barnacles, are sometimes found stranded on the beach, and on one occasion a very large squared log of mahogany was driven on the south shore, perforated on every side, to the depth of about six inches, by the Teredo navalis. Other instances might be quoted of the productions of distant countries being drifted to these islands.*

That the cedars of Bermuda, the palmetto, and, indeed, all the original vegetable productions of the group, were thus introduced by seeds or plants which had floated from the shores of America, there can be little doubt.

The direction of the winds and currents between the Northern States of America and the Bermudas, during the first four months of the year, is clearly demonstrated by the following occurrence: on the 4th January, 1853, the American brig, or brigantine, "Markland," laden with pitch pine lumber, for the port of Boston, was capsized and abandoned off that part of the American coast. On or about the first of May following, this "direlict" drifted within sight of the Bermudas, and was towed into the port of Hamilton, water-

---

* Sapindus saponaria, the common soapberry tree of the West Indies, growing in the Governor's grounds at Mount Langton, was raised from seed found on the south shore.—*See Williams' History of Bermuda.*

logged, and with only her foremast and bowsprit standing; having been four months buffeting with the fearful winter gales of those latitudes.

Robertson, in his History of America, states, that previous to the discovery of that portion of the globe, trees, torn up by the roots, were not unfrequently driven upon the coast of the Azores after a course of westerly wind, and that, "at one time, the dead bodies of two men with singular features, resembling neither the inhabitants of Europe nor of Africa, were cast ashore there." If we may venture to givé credit to this statement, it must be inferred, that these unfortunate men were natives of North America, who had been drifted to sea in their frail canoe of birch bark. It must be confessed there is difficulty in understanding how men placed in such desperate circumstances, would have supported life during the greater part of their perilous voyage, for they must have perished at no great distance from the Azores, to have been found in the state described by Robertson.

It is true that the prevailing westerly winds and the current of the gulf stream in that latitude, both greatly assist in drifting trees and other floating substances towards the shores of the Azores, but it is doubtful whether man, under the most favorable circumstances, could survive that slow process in an Indian canoe. By such means however, the distant islands of the Pacific appear to have been peopled, and were it not for the tempestuous character of the Atlantic, the islands of Bermuda would probably have been found populated by a race of Indians from the coast of North America.*

* The Island of Madeira, which is in the same latitude as the Bermudas, was also uninhabited by the human race when discovered by Bartholomew Perestrello.

The vignette below represents Hermitage, Smith's Parish, Bermuda, the residence of the late Gilbert Salton, Esq., for many years Collector of Customs for those Islands, taken from a sketch by Miss Ella Tucker.

GENERAL INDEX.

SCIENTIFIC INDEX.

Notes on the occurrence of new species, or, in fact, any communications relating to the Natural History of the Bermudas, will always be most thankfully received, and gratefully acknowledged; it being the author's intention (D.V.), to issue a second series of this work, should sufficient matter, and public interest, warrant the undertaking.

Please address—J. M. Jones, Esq., 1, Essex Court, Middle Temple, London.

London:—Printed by Bowden & Brawn, Princes Street, Little Queen Street. W.C.

Made in the USA
Monee, IL
04 November 2026

67385000R00125